The Military Potential of

CHINA'S

Commercial Technology

Roger Cliff

Prepared for the United States Air Force
Project AIR FORCE • RAND

The research reported here was sponsored by the United States Air Force under Contract F49642-96-C-0001. Further information may be obtained from the Strategic Planning Division, Directorate of Plans, Hq USAF.

Library of Congress Cataloging-in-Publication Data

Cliff, Roger.
    The military potential of China's commercial technology / Roger Cliff.
      p. cm.
    Includes bibliographical references.
    "MR-1292-AF."
    ISBN 0-8330-2939-8
      1. China—Defenses. 2. China—Armed Forces. 3. Industrial mobilization—
China. 4. China—Economic conditions—1976– I. Title.

UA835 .C57 2001
355'.07'0951—dc21

                                                00-068328

RAND is a nonprofit institution that helps improve policy and decisionmaking through research and analysis. RAND® is a registered trademark. RAND's publications do not necessarily reflect the opinions or policies of its research sponsors.

Cover design by Tanya Maiboroda

Published 2001 by RAND
1700 Main Street, P.O. Box 2138, Santa Monica, CA 90407-2138
1200 South Hayes Street, Arlington, VA 22202-5050
RAND URL: http://www.rand.org/
To order RAND documents or to obtain additional information, contact Distribution Services: Telephone: (310) 451-7002; Fax: (310) 451-6915; Internet: order@rand.org

This report examines the potential of China's civilian industry to serve as a source of advanced technology for China's military. It looks at the current standing of China's commercial technology in a number of industries with potential military applications and assesses prospects for further progress over the next 10 to 20 years. It should be of interest to researchers and policymakers who wish to know about China's potential for developing advanced military systems, as well as those who wish to know about China's commercial technological capabilities.

This study is part of a larger, multiyear project on "Chinese Defense Modernization and Its Implications for the U.S. Air Force." Other RAND reports from this project include

- Mark Burles, *Chinese Policy Toward Russia and the Central Asian Republics*, MR-1045-AF, 1999.

- Daniel Byman and Roger Cliff, *China's Arms Sales: Motivations and Implications*, MR-1119-AF, 1999.

- Zalmay Khalilzad, Abram N. Shulsky, Daniel L. Byman, Roger Cliff, David T. Orletsky, David Shlapak, and Ashley Tellis, *The United States and a Rising China*, MR-1082-AF, 1999.

- Mark Burles and Abram N. Shulsky, *Patterns in China's Use of Force: Evidence from History and Doctrinal Writings*, MR-1160-AF, 2000.

- Michael D. Swaine and Ashley Tellis, *Interpreting China's Grand Strategy*, MR-1121-AF, 2000.

This project is conducted in the Strategy and Doctrine Program of Project AIR FORCE under the sponsorship of the Deputy Chief of Staff for Air and Space Operations, U.S. Air Force (AF/XO). Comments are welcome and may be directed either to the author or to the project leader, Dr. Zalmay Khalilzad. The cutoff date for research in this report was November 2000.

PROJECT AIR FORCE

Project AIR FORCE, a division of RAND, is the Air Force federally funded research and development center (FFRDC) for studies and analyses. It provides the Air Force with independent analyses of policy alternatives affecting the development, employment, combat readiness, and support of current and future aerospace forces. Research is performed in four programs: Aerospace Force Development; Manpower, Personnel, and Training; Resource Management; and Strategy and Doctrine.

CONTENTS

FIGURES

If China's economy grows at expected rates over the next 20 years, by 2020 it will be larger than that of the United States (in purchasing power terms), which suggests that China has the economic potential to become a U.S. military rival. For China to actually become such a rival, however, China's defense industries would have to be able to produce weaponry technologically comparable to that of the United States. This would represent a dramatic advance over China's current military technology, which is still largely based on 1950s-era Soviet technology. Along with internal efforts to improve its technological capabilities, China's defense industry can draw on three external sources of advanced technology: direct transfers of military technology from abroad, purchasing from world commercial markets advanced components and equipment that China's defense industries are incapable of manufacturing, and technology diffusion from China's civilian industries. China is likely to exploit all three of these technology sources to varying degrees, but given limitations imposed by foreign governments on the first two—in contrast to the openness of China's civilian industry to foreign technology and investment—the third source could, in the long run, be the most promising source of knowledge and capability for China's defense industries.

CURRENT CAPABILITIES

Eight major civilian industries have the most potential for supporting military technology development: microelectronics, computers, telecommunications equipment, nuclear power, biotechnology, chemicals, aviation, and space. China has capabilities in all of these areas and has facilities in some of them that are quite advanced. These usually depend on imported components and machinery, however, and China's technological capabilities are overall well behind world standards.

In *microelectronics*, China's most advanced facilities are six to eight years behind the state of the art. At its present rate of progress, China would catch up to the state of the art by around 2008, but its advanced facilities are limited in number and depend on imported equipment. Until China succeeds in deepening its electronics capabilities, it will remain dependent on imports to keep up with advances in microelectronics technology.

China's capabilities for assembling low-end personal *computers* (PCs) are comparable to those of the advanced industrial nations, but its PCs are composed primarily of imported parts. Moreover, China has limited supercomputer capabilities. China does have large numbers of software professionals, but development of commercial software in China has so far been slow.

With regard to *telecommunications equipment*, China's ability to produce fixed-line switching systems has improved rapidly in recent years, with a market that was once dominated by foreign producers now split between Chinese and foreign companies. China produces fiber-optic cable but depends on foreign firms for advanced transmission technologies, such as synchronous digital hierarchy (SDH). China also possesses microwave transmission technology, including digital transmission systems, but it will need at least a decade to catch up with Western technological levels. Its ability to produce sophisticated terminal node equipment, such as cellular handsets, is limited, and the Chinese market is dominated by such foreign suppliers as Nokia, Ericsson, and Motorola. China also has limited communications satellite capabilities.

China's *nuclear power* industry is rudimentary. China has developed power plants for nuclear submarines, and it designed and built the nuclear power plant at Qinshan. But many of the critical components were imported, reflecting deficiencies in China's technological capabilities in this area. Nonetheless, Chinese capabilities are improving, and China will produce the majority of equipment for several foreign-led nuclear power plant projects now under way.

Biotechnology in China presents a dual picture. While basic science capabilities are strong and China is well endowed with highly trained scientists, the commercial biotechnology sector is small and China is weak in production technology. The fact that intellectual property rights are poorly protected in China further inhibits the growth of this industry.

As with its biotechnology, China's *chemical technology* is also considered stronger in basic research and development (R&D) than in production processes. A number of major Western countries have licensed technologies developed in Chinese laboratories. Chinese firms have been unable, however, to turn research results into commercial products, and China depends on imports for many chemicals.

China's *aviation* technology is mostly based on 1950s-era Soviet systems, and China does not yet produce an indigenous jet transport. A number of Chinese firms, however, many of which produce combat aircraft, are involved in component co-production arrangements with Western aircraft manufacturers. These arrangements are leading to rapid improvements in China's aviation technology.

As for China's *space* industry, its launch capability is impressive for a developing country, although the failure rate is somewhat high. China's satellite capabilities are limited, but include communications, photo-reconnaissance, meteorological, remote sensing, and experimental satellites. China also has a manned space program whose goal is to put astronauts in space by 2002.

PROSPECTS FOR FUTURE PROGRESS

Four characteristics affect a country's ability to acquire or develop new technologies: capabilities, effort, incentives, and institutions. *Capabilities* refers to the facilities, equipment, and trained personnel needed for technological innovation. *Effort* refers to the degree to which these capabilities are employed in the development of new technologies. *Incentives* refers to the macroeconomic environment, competition, and factor (capital, labor, and technology) markets.

Institutions refers to the available legal, industrial, training, and technology institutions. China's prospects for future technological progress can be evaluated by assessment along each of these dimensions.

With regard to *capabilities*, China's facilities and equipment for R&D do not appear to be optimal. Most facilities tend to be quite backward; others possess advanced equipment but do not fully utilize it. China's human capital base, as measured by formal education statistics, is adequate for a developing country but has important weaknesses. Total public expenditures on education are low: about 2.5 percent of gross national product (GNP) compared to 4 to 5 percent in countries such as India, South Korea, and Taiwan. Primary and lower-secondary education rates are good, however, with close to 90 percent of the population receiving a primary education and 65 percent receiving a middle school education in recent years (comparable to Taiwan in the 1970s). High school and college education rates, conversely, are low. Only about 20 percent of the population receives a high school education, and only 4 percent attend college, half attending three-year technical schools (comparable to Taiwan in the 1960s). India sends twice as high a proportion of its population to four-year universities as China does, while Japan, South Korea, and Taiwan send 15 to 20 times as high a proportion.

In technological development activities, however, absolute numbers of scientists and engineers may be more important than numbers as a proportion of the total population, and in this regard China compares more favorably with other countries. China awards roughly the same number of bachelor's degrees in science and engineering each year as India and the United States do, significantly more than Japan does, and far more than South Korea and Taiwan do. The United States, Japan, and India award many more advanced degrees in science and engineering than China does, however, although South Korea and Taiwan award many fewer. Overall, the numbers of scientists and engineers being trained in China appear comparable to those in Japan, lower than those in the United States and India, but far greater than those in South Korea and Taiwan.

The second characteristic affecting China's ability to improve its technology, *effort*, may be measured by numbers of scientists and engineers engaged in R&D, expenditures on R&D, numbers of scientific and technical publications, and numbers of innovations and patents. China has fewer scientists and engineers engaged in R&D than Japan and the United States have, but far more than India, South Korea, and Taiwan. Similarly, China's expenditures on R&D are much smaller than those of the United States or Japan, but much greater than those of India, South Korea, and Taiwan. As a proportion of gross domestic product (GDP), however, China spends less on R&D than any of those countries; its current rates are comparable to those of South Korea or Taiwan in the 1970s. Nonetheless, much of China's technological progress is the result not of indigenous R&D but of the technology transfers associated with foreign investment, and China has been receiving foreign investment in amounts unprecedented for a developing country.

China's output of scientific and technical publications in international journals is far less than that of the United States and Japan, but significantly greater than that of India, and far greater than that of South Korea and Taiwan. The proportion of patents granted to domestic applicants in China is less than that in South Korea and Taiwan in the 1980s, but greater than that in India.

As for the third characteristic, *incentives*, those provided by China's macroeconomic environment are mixed. High domestic growth rates and relatively stable exchange rates tend to increase the demand for innovation, but uncertainty about the domestic economy, fluctuating inflation rates, scarce credit and foreign exchange, and periodic political instability tend to discourage investment in technology development. Competition also provides mixed incentives. China overprotects many industries, although it does allow foreign competition in others, particularly those in which domestic capabilities are weak. China's industries have a strong incentive to compete on export markets, which tends to encourage technological progress, but many domestic industries are highly fragmented, with few firms achieving the scale to support significant R&D activities. Competition among the *providers* (as opposed to the consumers) of technology has also significantly increased in recent years, although some R&D institutes still depend on state-provided funds. Factor markets generally provide relatively poor incentives for innovation. The capital and technology markets are underdeveloped, although the labor market operates fairly efficiently, at least in technologically dynamic sectors.

Finally, the fourth characteristic for improving technology, China's *institutional structure*, is underdeveloped. The legal system is poorly developed, and intellectual property rights are often violated. Interfirm linkages have traditionally been weak, as have linkages between R&D institutes and productive enterprises. The government has, however, sponsored a number of programs intended to improve the institutional environment for technological progress in China. More important, many of the R&D institutes that formerly relied on state-provided funds are now remaining solvent by filling many of the institutional niches that were neglected under the central planning system.

In summary, China's prospects for technological progress are mixed. In terms of its capabilities (facilities, equipment, and human resources) and technological effort, China compares unfavorably to the United States and Japan and, on a proportionate basis, to South Korea and Taiwan. And its incentive and institutional structures for technological progress are also imperfect. By many measures, however, China's resources for technological progress look comparable to those of South Korea or Taiwan in the 1970s. Moreover, in terms such as absolute numbers of scientists and engineers and total spending on R&D, China already vastly surpasses smaller countries like South Korea and Taiwan. Thus, average technological levels in China in 2020 could be comparable to those in South Korea and Taiwan today, but, because of its greater size, China would possess state-of-the-art capabilities in many more areas than South Korea and Taiwan currently do.

IMPLICATIONS

China's overall technological capabilities will still significantly lag those of the United States by 2020. Capabilities in militarily significant areas may be somewhat stronger than average technological levels, but the long development times for military systems mean that the weapons China deploys in 2020 will largely reflect the technologies available in 2010 or earlier. Even though the Chinese military will not be the technological equal of the U.S. military by 2020, however, its technology will be substantially improved compared to its present technology. Moreover, potential theaters of conflict between China and the United States in the future may constrain the forces and capabilities the United States is able to employ. The U.S. military, including the U.S. Air Force, must prepare for the possibility of conflict under such conditions with a Chinese military that by 2020 will be significantly more advanced than it is at present.

ACKNOWLEDGMENTS

This study would have been impossible without the contributions of a number of people. In particular the author would like to thank Anthony Lanyi and the IRIS Center at the University of Maryland, along with the Office of Net Assessment in the U.S. Department of Defense, for sponsoring a paper that became the stimulus for this more detailed study; Monica Pinto for providing him with a copy of her own study of China's science and technology capabilities along with invaluable materials that were used in this study; Greg Felker for directing him to the theoretical framework used in Chapter Four; Rob Mullins for valuable insights about the nature of technological progress; Toshi Yoshihara for research assistance including an initial draft of the section on China's space technology; Andrew Mok for research assistance and suggestions on data presentation; Kevin Pollpeter for research assistance; Erica Downs for acquiring several hard-to-find documents; Jennifer Casey and Gail Kouril of the RAND library for assistance in locating and acquiring many of the materials used in this study; Barry Naughton for useful comments on the IRIS paper; Phyllis Gilmore for editorial guidance; Susan Spindel for turning the draft text into a RAND document; Erik Baark and Richard Suttmeier for rigorous and insightful reviews of an early draft; Richard Neu and Zalmay Khalilzad for useful comments and overall guidance and support; Andrew Hoehn and Barry Pavel for allowing me the time off to complete this report; and Jeanne Heller and the other staff of RAND's publications department for superb editorial support. Any errors in this report are the sole responsibility of the author.

Chapter One

INTRODUCTION

China's huge size and recent rapid economic growth suggest that it has the potential to become a superpower rival to the United States in the early 21st century. According to the World Bank, China already had the world's second largest economy in purchasing power terms in 1995, and its economy was projected to grow at an annual rate of about 6 percent between 1995 and 2020 (World Bank 1997b, pp. 6–9, 129, 134–136; World Bank 1997a, p. 20).[1] If the U.S. economy grows at an average annual rate of 3 percent over the same period, China's economy will surpass it in size before 2020.[2]

Such an economy would provide China with the economic base for fielding a military comparable to that of the United States. However, this does not mean that China's defense industries would be technologically capable of equipping a military that could challenge the United States. China's defense manufacturers are currently quite backward. The major combat platforms (aircraft, naval vessels, armored vehicles) they produce are still largely based on 1950s-era Soviet designs. Pockets of respectability exist in some areas, such as short-range ballistic missiles and anti-shipping cruise missiles, but in

---

[1]According to the U.S. Arms Control and Disarmament Agency (1997, pp. 65, 91), China's economy is already about as large as the Soviet Union's was at its peak in the 1980s. The validity of such estimates is, of course, arguable. More important, China's leaders clearly do not intend to devote as great a proportion of their economy to military spending as the Soviets did, recognizing that short-term military strength would come at the expense of long-term economic growth.

[2]Some projections are even more dramatic. A 1995 RAND study (Wolf et al. 1995) estimated that the size of China's economy would equal that of the United States by 2006.

general China possesses none of the high-technology weaponry demonstrated so effectively by U.S. forces during Operation Desert Storm (Arnett 1994, pp. 359–383). For China to field a military that is technologically capable of challenging the U.S. military by 2020, it would have to make major improvements in the technological capabilities of its defense industries.

An alternative approach, of course, would be for China, like many countries, to rely on imported weapons. Indeed, it appears that China is increasingly resorting to imports, particularly from Russia, for its weapon needs (e.g., see Brodie 1999, p. 13; Pomfret 2000, pp. A17–A18). In the long run, however, China is unlikely to depend on imports as its main source of weaponry, for several reasons. First, Chinese leaders are uncomfortable with relying on foreign suppliers in this area. Twice in its recent history, China has suffered a sudden cutoff of imported military technology—first in 1960 when the Soviets withdrew their technical aid, and again in 1989 when Western nations suspended arms sales in the wake of the Tiananmen incident. Second, other countries, out of concern for their own security, are unlikely to provide China with a full range of advanced weapons in the quantities China would need to become a world power. Third, the expense would be prohibitive. The United States spent $55 billion on military procurement in fiscal year 2000. Although Russian equipment comes at a somewhat lower price than Western equivalents, for China to equip its forces with imported weaponry even half as well as the United States does would require roughly a doubling of China's overall defense budget and huge amounts of foreign exchange. Finally (and perhaps most important of all), Chinese leaders are unlikely to rely on imported weapons because doing so would be inconsistent with their vision of China as a sophisticated, self-reliant world power. (Cohen 2000, p. B-1; IISS 1995, p. 275; Arnett 1994, p. 361; Sun 1991, p. 173.)

One potential source of technology for upgrading the capabilities of China's defense industries is China's civilian economy. Since the late 1970s, China's economy has been increasingly open to foreign trade and investment, which has led to an unprecedented amount of capital and technology pouring into China, particularly since the early 1990s. As a result, the products flowing out of China are of growing sophistication. Once confined to textiles and handicrafts, China's exports are now increasingly dominated by consumer electronics and computer components.[3] This suggests that China's civilian economy is rapidly upgrading and could represent a source of technology for China's defense industries. Indeed, at least one analyst has asserted that "in a few years, China will [be able to] quickly translate civilian technological power into its military equivalent" (Cohen 1996, p. 51).

The coming years may represent a particularly favorable time for China to acquire advanced military technology. Many analysts argue that a "revolution in military affairs" (RMA) is under way and that it will render obsolete current military technology that, while vastly more effective than what was employed in World War II, has changed little in qualitative terms since that time. Just as the nature of warfare changed fundamentally between World War I and World War II, so too might the current RMA fundamentally alter the nature of warfare in the 21st century. If so, the currently overwhelming U.S. advantage in modern weaponry may be irrelevant in the new era. For a country such as China to challenge the United States militarily, however, it would have to possess the technological capability needed to produce the new generation of weaponry.

REPORT STRUCTURE

The study reported here investigates the degree to which China's commercial technology could contribute to the improvement of its military technology. Chapter Two places China's civilian industry in context with other potential sources of technology for China's defense industries. Chapter Three documents the current technological capabilities of a number of Chinese industries with potential military applications. Chapter Four assesses China's prospects for further technological progress. Chapter Five, the conclusion, examines the implications of the current and future capabilities of China's commercial technology vis-à-vis China's ability to develop military systems that could present serious challenges to the U.S. military.

---

[3]Of China's $184 billion in exports in 1998, 36 percent consisted of "mechanical and electrical products," as opposed to 22 percent for "textile materials and products" (National Bureau of Statistics 1999, pp. 577, 582, 588).

METHODOLOGY

The two main parts of this study, described in Chapters Three and Four, respectively, employ different methodologies. The first, evaluation of China's current technological capabilities, relies to a large extent on descriptions of China's industrial technology in Western (American or British) trade journals. These reports were identified through extensive searches of several periodical indexes and are of value because of the comparative perspective they provide on China's technological capabilities. Lesser use was made of accounts in the Chinese press, as these tend to focus on touting China's technological achievements or on particular breakthroughs that have resulted in China becoming a "world leader" in specific technology areas. Not only is the accuracy of some of these claims suspect, but most of the claimed accomplishments are in highly specialized technologies whose commercial or military significance is unclear. Chinese sources have been used as an important source for statistics and other types of factual information.[4]

The second main part of the study, the assessment of China's potential for future technological progress, relies largely on statistical data on measures such as education levels, overall R&D funding, numbers of scientists and engineering in R&D, and so on. For China, this information was largely obtained from official sources such as the *China Statistical Yearbook* and *China Science and Technology Indicators*; comparative data for other countries came from similar sources in those countries. The data on China refer to China as a whole, not just to its defense industries. They therefore are not particularly sensitive, and there is no reason to suspect that they are disguised or distorted. Chinese aggregate statistics are assumed to be generally reliable.

One last point is appropriate here. A large number of sources were consulted for these analyses. For readability reasons, all references drawn on by a particular paragraph are generally collected at the end

---

[4]An extensive study of Chinese trade and technical journals, as opposed to press reports, would no doubt yield much useful objective data about specific technology areas. This study focused instead on providing an overview of overall capabilities in several broad industries.

of the paragraph. The exceptions apply in cases where specific attri-
bution for a particular claim is required.

Chapter Two

BACKGROUND

China's defense industries have two avenues open for improving their technological capabilities: technology development efforts within the defense sector and taking advantage of technologies available outside the defense sector. The latter, external, approach can be done in three main ways: through direct transfers of military technology from abroad, through purchasing advanced components and equipment from world commercial markets, and through technology diffusion from China's civilian industries.

Direct transfers of military technology from abroad entail foreign countries simply providing China's defense industries with the specialized equipment, components, and know-how needed to produce advanced weapon systems. Much of China's defense industry is, in fact, the result of the Soviet Union's having provided such assistance in the 1950s. China's defense industries currently do receive some direct assistance from Israel and Russia, but these technology transfers have been limited. Israel's assistance has primarily been confined to subsystems, such as radars and fire control systems, rather than involving complete weapon systems. Russia has agreed to allow co-production of fighter aircraft, but Russian manufacturers refuse to share the most critical technologies (such as engine manufacturing). An unknown number of scientists and engineers formerly employed by Soviet defense industries are, however, reportedly now working in China.

The second potential external source of technology for China's defense industries consists of imported equipment and components. That is, it may be possible for China's defense industries to significantly improve the capabilities of the weapons they produce by simply purchasing from abroad critical components or equipment that China is incapable of manufacturing domestically. Such an approach has been applied successfully in certain civilian technologies[1] and has been attempted to some degree by China's defense industries. The potential of this approach is limited, however, by Western export controls on military and "dual-use" technology.

The third possible external source of advanced technology for China's defense industries is China's civilian industries. China's commercial enterprises could serve as a source of know-how, equipment, and components for improving the technological capabilities of China's defense industries. The Chinese civilian sector has been open to foreign investment and technology transfer for over 20 years now, and although many foreign investments were initially low-technology, labor-intensive enterprises, the technological level of foreign joint ventures has recently been steadily increasing. Weapon system production generally involves equipment and components developed for purely military applications, but in many cases these items are produced from equipment and components that are inherently dual-use. Thus, much military production ultimately rests on a civilian technological base. Particularly in recent years, the effectiveness of modern weapon systems is largely based on advances that ultimately trace back to technologies having primarily commercial, rather than military, applications (Cohen 1996). More generally, the reservoir of knowledge and capability that exists in China's civilian industries could be tapped to improve China's defense industries.

China's future ability to produce advanced weapon systems will undoubtedly be the result of all three of these technology sources as well as internal R&D efforts, with no one of them playing an exclusive role. Domestically produced weapon systems are likely to be the results of both indigenous R&D efforts and foreign technology transfers and will incorporate both imported components and the products of China's domestic industries. However, if we assume that Western export controls will continue to prevent China from simply import-

---

[1]See, for example, the description of the development of the HJD-04 telephone switching system in Shen 1999, pp. 105–141.

ing the components and equipment it needs to produce advanced weapon systems and that Russian and Israeli assistance to China will remain limited, then China's domestic technological capabilities (i.e., those of both its civilian and its defense industries) will, ultimately, be the determining variable in China's ability to produce such systems. Moreover, the globalization of investment and technology trade in recent decades has given China's civilian industry access to a wide range of advanced commercial technologies, which means that civilian industry could well be a valuable source of technology for China's defense sector. A massive allocation of resources to priority defense programs could certainly substitute for any deficiencies in the broader civilian technology base, but the experiences both of China during the 1960s and 1970s and of the Soviet Union throughout the Cold War suggest that such efforts are extremely costly and probably unsustainable over the long run. Efforts to improve the technological capabilities of China's defense sector will obviously be the ultimate determinant of whether China acquires the capacity to indigenously produce advanced weapon systems, but China's civilian industry represents an important source of technology for these efforts.[2]

---

[2]Unless foreigners are willing to transfer a complete range of military technologies, the defense sector's internal R&D efforts will be crucial. No matter how advanced the technological level of China's civilian industries or how sophisticated the civilian equipment and components available for import, all Chinese weapon systems ultimately have to incorporate purely military technologies, and these have to be developed indigenously. In other words, even if China's civilian industries possessed all the constituent technologies needed to produce advanced military systems, additional research would be required to develop and test weapon systems based on these technologies. Similarly, even if China could simply import the equipment and components needed to produce advanced weapon systems, additional R&D would be required to determine how to use them to produce those systems. Thus, the technological capabilities of China's civilian industries should be viewed as important contributors to, but not the sole determinants of, China's future military capabilities.

Chapter Three

CHINA'S CURRENT CIVILIAN TECHNOLOGY

Modern military technology is closely related to civilian technology. Of the 18 broad technology areas included in the U.S. Department of Defense's *Militarily Critical Technologies List* (OUSD(A&T) 1996), most correspond directly to major civilian industries, with the remainder described as "supported" by technology areas that correspond directly to major civilian industries. This chapter examines China's technological capabilities in the eight technology areas in the *Militarily Critical Technologies List* that correspond most closely with major civilian industries: microelectronics, computers, telecommunications equipment, nuclear power, biotechnology, chemicals, aviation, and space.

MICROELECTRONICS

The centrality of electronics to modern military capabilities is widely recognized,[1] and, especially since Operation Desert Storm so dramatically underscored this importance, the development of China's electronics industry has been accorded a high priority by China's leadership.[2] Electronics was declared a "pillar industry" in China's Ninth Five-Year Plan (1996–2000), and 590 billion yuan (about $70 billion) was allocated to upgrade China's electronics technology

---

[1]According to the *Militarily Critical Technologies List*, "Electronic devices and components contribute . . . perhaps more so than any other technology, to the current technological edge of most U.S. military systems" (OUSD(A&T) 1996, p. 5-1).

[2]Project 909 (see below) is said to have the "emphasis of emphases," equivalent to China's atomic bomb project in the early 1960s (*Solid State Technology* 1996, pp. 50, 54).

base within the Plan period.[3] Even before the electronics sector was accorded this special status, its output had been growing at more than 20 percent a year since 1981, double the overall growth rate of the economy, and it has grown at a rate of 27 percent a year since 1993. Nevertheless, China's electronics technology still significantly lags that of the industrial nations and remains highly dependent on imported equipment and know-how. (Hu 1997, p. 1; Simon 1992, p. 22; Simon 1996, pp. 8, 9; Zhang 1996, p, 22; Huchet 1997.)

The critical technology in microelectronics is the ability to produce integrated circuits (ICs). A measure of the limitation of China's capabilities in this regard can be found in the fact that China produced only 560 million of the 5.7 billion ICs it consumed in 1995. This was still a significant improvement over just three years earlier, however, when only 97 million ICs were produced and China's IC production capability was assessed as "extremely low and limited to ICs used in consumer goods, such as televisions and refrigerators" (Hui and McKown 1993, pp. 18–19).[4] (Simon 1996, p. 9; Simon 1992, p. 24.)

As of the late 1990s, the most advanced IC production facility in China was a joint venture between NEC and the Shougang Iron and Steel Corporation that utilized 6-inch, 0.8 micron technology to produce 4 Mb dynamic random access memory (DRAM) chips. Plans for even more-advanced facilities included Motorola's wholly owned IC fabrication facility in Tianjin, which was expected to be producing 8-inch chips with linewidths of 0.5 microns, and Project 909, a State Council–sponsored $1 billion joint venture between NEC and Hua Hong Microelectronics that also planned to introduce 0.5 micron technology. By this time, however, Western manufacturers had advanced to 0.18 micron technology, and IC manufacturing technology

---

[3]China's electronics industry may be divided into four main subsectors: consumer electronics, microelectronics, computers, and telecommunications equipment. Much of China's electronics production consists of consumer electronics with little military significance, such as air conditioners, refrigerators, color televisions, videocassette recorders, and electric fans. The other three sectors, however, all have military significance. This section discusses microelectronics; computers and telecommunications equipment are examined in the following two sections.

[4]Even though production increased fivefold from 1990 to 1995, demand increased fifteenfold. Thus, as a *proportion* of demand, domestic production actually fell from 30 percent in 1990 to 10 percent in 1995 (Simon 1996, p. 9).

in China was six to eight years behind the world state of the art. This represented an improvement over the late 1980s, when China was at least 12 years behind, however, and if past trends continue, the most advanced producers in China will catch up to world state of the art by about 2008. (Tilley and Williams 1997, pp. 147–148, 150; Simon 1996, pp. 15–16; Erkanat and Fasca 1997, pp. 48, 52, 142; *Solid State Technology* 1996, pp. 50, 54; Hu 1997, p. 1; Lu 1999; Schoenberger 1996, p. 124.)

Despite these advances, important limitations exist in China's IC capabilities. The most important problem is that China lacks the capability to manufacture the lithography tools ("wafer steppers") used in producing ICs. Thus, while China may acquire facilities capable of manufacturing state-of-the-art chips, these facilities will continue to depend on imported machinery. The costs associated with importing this equipment mean that only a small number of facilities in China will be producing at state of the art, and as the state of the art advances, even China's most advanced facilities will soon be second rate unless China continues to import new production equipment. (Geppert 1995, p. 45; Huchet 1997.)

More generally, China's electronics industry has had difficulty in absorbing foreign technology and has been unable to convert its own laboratory breakthroughs into improvements in production. China's electronics sector suffers from poor infrastructure and an immature peripherals industry. Turf battles between the State Planning Council and the Ministry of Finance and between the Ministry of Electronics Industry and the Ministry of Posts and Telecommunications (the latter two of which have since been recombined into a single ministry) have also limited China's ability to acquire imported technology. An additional barrier to acquiring advanced foreign technology is (ironically) the weakness of protections for intellectual property rights in China, as foreign electronics firms are unwilling to transfer technology to China without assurances that they will retain control over it. (Simon 1992, p. 24; Simon 1996, pp. 9, 15–16; *Solid State Technology* 1996, p. 54; Erkanat and Fasca 1997, pp. 1, 37, 48, 52, 142.)

Nonetheless, the Chinese government has high hopes for the electronics sector and will continue to push efforts to upgrade technological levels. The supply base for essential components and subassemblies is improving, and China has almost caught up to world levels in its technology for board-level assembly of final products. Although China will not join the ranks of the world leaders in IC technology any time soon, one industry executive states that "there is little doubt in my mind that in twenty years the Chinese [electronics] industry will be formidable" (Erkanat and Fasca 1997, p. 48). (Simon 1996, p. 9; Tilley and Williams 1996, p. 335; Tilley and Williams 1997, pp. 145, 150.)

COMPUTERS

Since a modern military cannot function without computers, both general purpose and specialized, it is important to be able to design and manufacture them. But the ability to design and manufacture computers is also important in that it reflects a more general capacity to design and develop devices employing microelectronics, such as guidance and electronic warfare systems. China's capabilities in this area have been gradually improving. In particular, China's ability to produce low-end personal computers (PCs) is essentially comparable to that of the advanced industrial nations. According to one American executive, Chinese-produced PCs are "very advanced systems and very competitive with multinationals" selling in China (Roberts 1997, p. 58). This reflects rapid progress over the past few years and is the outcome of a change in official policy—from focusing on high-end mainframes and minicomputers and trying to develop a completely indigenous capability to produce computers during the early 1980s, to focusing on low-end products incorporating imported components. The results have been impressive. In 1990, most of the 600,000 PCs in use in China had been imported; by 1995, annual domestic production exceeded 800,000 units. By 1998, domestic production was 2.9 million units. Overall, computer hardware production grew at a 29 percent annual rate from 1987 to 1993. As noted in the preceding section, however, China depends heavily on imports for the ICs used in these machines, and, altogether, less than 25 percent of the components used in the PCs produced in 1995 were locally sourced. Computers capable of more than a certain number of calculations per second are subject to export controls in the West, but Chinese scientists have apparently achieved some success in developing their own high-speed systems using clustered workstations and parallel processors.[5] (Hui and McKown 1993, pp. 14–16, 18; Kraemer and Dedrick 1995, pp. 64, 66–67, 69; Gan and Zhang 1992, p. 113; State Statistical Bureau 1996, p. 442; Simon 1996, p. 13; Erkanat and Fasca 1997, p. 37; Hu 1997, p. 1; Xiao 1993, p. 42; Li 1996, p. 12; Sokolski 1999; Suttmeier 1997, p. 317; National Bureau of Statistics 1999, p. 446.)

China's software capabilities are just beginning to develop. Traditionally, Chinese computer users have relied largely on pirated software, but software development, drawing on a large pool of computer programmers, is emerging as an area of future strength in China. As of 1993, China already had more software professionals than any other country besides the United States. Chinese firms sold only about $50 million of packaged software in 1996, but this was expected to have increased to as much as $1 billion in 2000. Overall sales of software and information services grew from about $90 million in 1990 to around $3 billion in 1997, a 65 percent annual growth rate. Capabilities in some areas are internationally competitive. Beijing Founder Electronics' Chinese-language publishing software dominates not only China's market, but those of Hong Kong, Taiwan, and Singapore as well. Unlike other industries, software requires relatively low levels of investment in equipment, and its relative newness means that a late-developing country such as China is starting at roughly the same place as the advanced nations. Moreover, language and cultural differences limit the utility of foreign-developed computer applications in China, providing a major market opportunity for Chinese software developers. According to one computer executive in 1999, "The difference between the United States and China in computer science was 10 years if you visited here 10 years ago, and it was shortened to three years if you visited five years ago. . . . It's less than one year, or almost on the same level, now" (Liu 1999). Piracy remains a major problem, however, for domestic firms as well as potential foreign investors. Until piracy is controlled, it will act as a damper on the development of the domes-

---

[5]The limits on the processing power of computers that may be exported to China have been adjusted over time in response to advances in the speed of widely available microprocessors. As of 13 October 2000, computers performing more than 28,000 million theoretical operations per second (MTOPS) could not be exported to China without an export license (Office of the Press Secretary 2000).

tic software industry. (Hui and McKown 1993, pp. 15, 16; Kraemer and Dedrick 1995, pp. 65, 70–71, 72; Saywell 1997, p. 60; Roberts 1996a, pp. 62–63; Lu 1999; Hilborn 1998, p. 50.)

TELECOMMUNICATIONS EQUIPMENT

Another sector of electronics significant to the military is telecommunications. Decision-makers' access to information is a crucial element of success in military conflict,[6] and, as the means for gathering information proliferate, the demands on telecommunications equipment to transmit that information to decision-makers increase proportionately. Military and civilian requirements for telecommunications equipment have been converging, and the development of civilian technology has been so rapid that the most–advanced technology is now embodied in civilian, rather than military, equipment. Thus, perhaps more so than any other area, civilian telecommunications technology does not just provide a jumping-off point for developing military telecommunications technology—in many cases, the two are identical. (Anthony 1996, pp. 557, 558; General Accounting Office 1996b, pp. 3–5.)

Despite the inherently dual-use nature of telecommunications technology, however, a country cannot simply purchase the equipment it needs from abroad and divert it for military use. Effective systems integration is crucial to the proper functioning of a telecommunications system, and companies that supply these systems are generally closely involved in the installation and implementation of the system. It is unlikely, therefore, that a supplier could be deceived regarding the nature of the system's end use, and all major suppliers of telecommunications equipment are from either North America or Western Europe, where they are subject to restrictions on exports to military users in China. Nonetheless, the skills learned in constructing and operating a civilian telecommunications system could potentially be applied to a military network, especially if the military is involved in the network's ownership, as is often the case in China. (Anthony 1996, pp. 552, 554, 555, 558, 559.)

---

[6]For example, see Johnson and Libicki 1995.

The primary components of a modern telecommunications network include switching systems, line transmission systems (copper wire and optical fiber), microwave radio networks, satellite communication systems (satellites and earth stations), and terminal node equipment (telephone sets, cellular telephones, pagers, computers, etc.). The technological requirements for deploying such a network include the abilities to manufacture the hardware, develop the software, and design and implement an integrated system. China has demonstrated partial capabilities in all of these areas. (Chen 1993, p. 21; Liu 1993, p. 38; Qiu 1993, p. 46; Shen 1999, p. 15.)

China's ability to manufacture switching systems has been improving rapidly. Foreign switching equipment producers began establishing manufacturing bases in China in the 1980s, and a total of seven foreign companies are now producing switching equipment in China. Recently, indigenously designed switching systems have begun to compete with both imports and systems produced by foreign subsidiaries. Two designs for medium-capacity systems (capable of handling 30,000 subscriber lines) had been completed by 1991, and high-capacity systems (capable of handling up to 100,000 subscriber lines) have recently been developed. This achievement is regarded as a "breakthrough" that has broken the "foreign monopoly" in this area (Hu 1997). Indigenously designed systems have been capturing a growing share of the domestic market. As recently as 1994, foreign manufacturers still dominated the market for switching systems in China, but by 1998, Chinese companies were said to have gained nearly 50 percent of the market.[7] These systems rely on imported components, particularly ICs, but those components are widely available on world markets.[8] (Rehak and Wang 1996, pp. 12, 13; Chen, Yan, and Li 1993, pp. 56, 59–60; Hu 1997; Denton 1996, p. 21; Granitsas 1998, p. 12; Shen 1999, pp. 105–141.)

---

[7]In 1993, only about a quarter of the 10 million lines of digital switching equipment that had been installed were provided on systems that had been manufactured in China, and most of these were produced by local subsidiaries of foreign companies. In 1994, less than 10 percent of the market went to purely domestic manufacturers. (Hao 1993, p. 55; Granitsas 1998, p. 12.)

[8]Another vital component consists of the switches themselves. Several Western companies have joint ventures producing switches in China including Sweden's Ericsson and Lucent Technologies, the latter of which is producing switches described as "the world's most advanced digital programme-controlled switch," at a joint venture in Qingdao. (*China Daily* 1997; Schoenberger 1996, p. 116.)

The medium of choice in line transmission systems is now fiber-optic cable. Currently, both domestically produced and imported optical fiber systems are in use in China, with the Chinese-produced systems said to perform up to international standards, and research is under way on technologies to increase transmission rates. Following the lifting of Western export restrictions in 1994, China gained access to synchronous digital hierarchy (SDH) technology, which enables high-speed transmission of data, video, and voice traffic (and will be central to future defense communications networks in the United States). Subsequently, most large fiber-optic equipment manufacturers have begun setting up joint ventures to produce SDH systems in China. (Ye and Ren 1993, pp. 48–50; Qiu 1993, p. 46; Zhou 1993, pp. 63–65; Rehak and Wang 1996, p. 12; General Accounting Office 1996b, pp. 3–4.)

China is said to have a mature capability in microwave transmission technology, although it is estimated that it will be at least a decade before China's equipment catches up to that of foreign manufacturers in terms of technical level and reliability. Microwave radio technology was initially developed in China without foreign assistance, and China began constructing analog microwave transmission systems during the early 1970s. Research on digital transmission systems began in the 1970s; by the late 1980s, high-capacity systems were being deployed. As of the early 1990s, most of China's microwave network was analog, but the goal was for the system to be completely digital by 2000. (Liu 1993, pp. 39–40; Yao, Cao, and Wang 1993, p. 43; Rehak and Wang 1996, p. 13.)

The technology for producing ordinary telephone sets does not present a challenge for China, but the technology for wireless equipment—such as cellular handsets, second-generation cordless telephones,[9] and pagers—is less advanced. Over 90 percent of the cellular handsets sold in China are manufactured by foreign companies, although some of these are assembled at joint ventures in China. Capabilities for pagers are stronger, and domestic firms have

---

[9]Second–generation cordless telephones (CT2), an alternative to conventional telephones, connect to the switching office via radio waves rather than wires. They are of limited range (about 500 meters) and therefore, unlike cellular telephones, are not truly mobile. But they represent an attractive alternative to conventional telephones in certain environments, such as high-density urban areas. (Denton 1996, pp. 18–21.)

developed their own Chinese-character paging systems. (Denton 1996, p. 21; Yao, Cao, and Wang 1993, p. 45; State Statistical Bureau 1996, pp. 442, 529; Elegant 1998, p. 11.)

China has demonstrated limited capabilities in developing satellite communications equipment. The first experimental communications satellite was launched in 1984, and three 4-transponder units were put in orbit between 1988 and 1990. An improved, 24-transponder model has been co-developed with Germany. The first launch failed in late 1994, however, due to problems with the satellite's attitude control system, and a replacement was not launched until May 1997. Even if the earlier launch had been successful, China would have experienced a shortage of transponder capacity (the shortfall has been made up by leasing transponders on, or purchasing, foreign-made satellites), indicating that China has encountered difficulties in mastering the technology for producing communications satellites. China is also mostly dependent on imported equipment (except for the antennas) for its satellite ground stations. (Clark 1997b, pp. 10–12; Yao, Cao, and Wang 1993, pp. 43–44.)

NUCLEAR POWER

The most obvious military application of nuclear power technology is in the production of nuclear weapons, but nuclear power plants also have important military uses in naval vessels and spacecraft. China, of course, already possesses nuclear weapons and has built nuclear power plants for submarines. The latter have not been entirely satisfactory, however, so improvements in China's civilian nuclear technology could lead to improvements in its military applications.[10] (Lewis and Xue 1994.)

China's nuclear power industry is still relatively undeveloped. Currently, two facilities are in operation: at Daya Bay in Guangdong

---

[10]In the United States, the close relationship between civilian nuclear power and defense programs is illustrated by the fact that Westinghouse, the United States' primary nuclear power company, also works on propulsion systems for the U.S. Navy and projects for the U.S. Department of Energy, which oversees the country's nuclear weapons (Hibbs 1998a, p. 3).

province and at Qinshan in Zhejiang province.[11] These facilities have a total capacity of 2100 MW, only about 1 percent of China's total electrical generating capacity. Four new plants are under construction, however. These joint ventures, involving assistance from French, Canadian, and Russian firms, have already led to substantial improvements in China's nuclear power technology. In addition, the 1998 lifting of the ban on U.S. nuclear cooperation with China means that U.S. technology will be available to China for future nuclear power projects. (Suttmeier and Evans 1996, pp. 16–17; *Nuclear Engineering International* 1993, pp. 18–21; Wei 1998; Williamson 1998, p. 9; Hibbs 1998b, pp. 13–14; Hibbs 1998c, pp. 14–15.)

China's first nuclear power facility was built at Daya Bay by France's Framatome and was constructed primarily using imported equipment. Qinshan, which began generation in 1991, is a Chinese-designed facility that draws on China's experience in producing reactors for submarines, but Chinese industry was limited at that time in its ability to manufacture critical components. Although 80 percent of the components were manufactured domestically, they were mostly the nonnuclear "balance of plant" equipment. The pressure vessel, primary coolant pumps, some instrumentation, and parts of the turbine were all imported. In addition, some of the domestically produced components performed so poorly that they had to be replaced. Also, the capacity of Qinshan's single 300 MW reactor was small compared to the 900 MW reactors installed at Daya Bay. (Suttmeier and Evans 1996, p. 18; *Nuclear Engineering International* 1993, p. 21; Wei 1998; Hibbs 1998b, pp. 13, 14.)

In addition to further contributing to the improvement of China's nuclear power technology, the four new plants currently under construction reflect advances that have already occurred. Qinshan Phase II, for example, will be powered by two 600 MW reactors designed and built by China when it is completed in 2002–03. At Framatome's new facility in Ling Ao, Guangdong province, Chinese companies are providing a number of components, including the steam generators and reheater units. Although Framatome turned down a Chinese firm's request to produce the reactor vessels for

_____

[11]China has also constructed a 300 MW nuclear power plant, based on its Qinshan design, in Pakistan (Wei 1998).

Ling Ao because it was concerned about quality assurance, that firm's capabilities have now improved to the point that it will be allowed to produce pressure vessels for any future Framatome facilities in China. Similarly, Westinghouse has said that it will allow Chinese firms to manufacture the turbine-generators, pressurizers, and condensers for any facility it is contracted to build in China. (Hibbs 1998b, p. 13; Hibbs 1998c, pp. 14–15.)

China has a number of additional indigenous R&D efforts related to nuclear power. Programs are under way on advanced light-water reactors, high-temperature gas-cooled reactors, fast breeder reactors, and nuclear fusion. There were also plans for a pilot reprocessing plant for extracting plutonium from spent fuel, to be constructed by the end of 2000. (Suttmeier and Evans 1996, pp. 18, 20–21; *Nuclear Engineering International* 1993, pp. 20–21, 22.)

BIOTECHNOLOGY

Biotechnology has obvious military implications as a means for developing biological weapons, but it is also important in providing defense against biological weapons in the form of detection, warning, and identification systems, and technologies for prophylaxis and therapy. Biotechnology may have military applications in other areas as well in the future, such as in nonlethal weapons (e.g., microbes that destroy fuel supplies). (OUSD(A&T) 1996, pp. 3–1.)

Despite government efforts dating back to as early as 1980, China's biotechnology industry is still small, with total sales of about $500 million in 1998. Substantial government and academic laboratory research has been sponsored, however, and biotechnology was one of the major research areas under the Chinese government's High-Technology Plan of 1986. China's leadership aims to achieve parity with the developed world in biotechnology research by 2005 and to become an important player in the industry. (Saywell 1998, p. 49; BioIndustry Association 1996, pp. 7–122; Baark 1991a, p. 87; *Futures* 1989, p. 227; Layman 1996, p. 13.)

There are a number of barriers, however, to the emergence of a dynamic biotechnology industry in China. One is the enforcement of patent rights. As is true for the software industry, inadequate protection of intellectual property rights discourages commercial research. A company that appropriates the research results of others incurs no research costs itself, and thus can undercut the original developer. A second problem is the lack of a strong market for sophisticated biotechnology products in China, which means that firms must either find markets abroad or receive government subsidies to sustain their operations. (BioIndustry Association 1996, pp. 11–12, 16–17; Layman 1996, p. 13; Baark 1991a, pp. 89–90; Saywell 1998, p. 49.)

Partly as a consequence of these conditions, China's biotechnology capabilities are stronger in basic science than in production technology. Most research efforts focus on the more glamorous "upstream" capabilities, such as recombinant DNA technologies, while neglecting key "downstream" technologies needed for the purification, formulation, and commercialization of products. And although many Chinese laboratories have good research capabilities and are well equipped, they lack management experience and the expertise needed for commercial production—such as the ability to scale up or an understanding of quality assurance, safety, and regulatory issues. Finally, as is often true in China, laboratories are eager to purchase sophisticated foreign equipment but are typically unwilling to spend foreign exchange on the maintenance and supplies needed to ensure the continued effective operation of that equipment. (BioIndustry Association 1996, pp. 11–12, 14; Baark 1991a, p. 88; Layman 1996, p. 13; Kinoshita 1995a, pp. 1147–1149.)

Nonetheless, China is viewed as having considerable future potential in biotechnology. Because the field is new and relatively undeveloped, the most important asset is first-rate scientists, an asset in which China is considered well endowed. China's abundant raw materials, huge storehouse of traditional herbal medicines, rich biodiversity, and relaxed regulatory environment are also regarded as advantages. According to one assessment, "All of the evidence points to an extremely rapid rate of growth in China possibly to place China on a par with the Western economies within 15 years" (BioIndustry Association 1996, p. 14). This judgment may be overly enthusiastic, but it does seem plausible that China could develop strong capabilities in a number of niches by that time. (Layman 1996, p. 13; *Technology Review* 1992, p. 19; *Futures* 1989, p. 227; Saywell 1998, p. 49.)

CHEMICALS

Like biotechnology, chemical technology also has obvious military implications, in this case in the form of chemical weapons. China is believed to already have an advanced chemical weapon capability, but chemical technology has military significance in a number of other areas, including the manufacture of explosives, missile propellants, electronics, and advanced materials. (Office of the Secretary of Defense 1997, p. 10; OUSD(A&T) 1996, p. 11-1.)

China has a huge chemical industry, with over 28,000 chemical-producing enterprises. In 1998, the gross output value of "raw chemical materials and chemical products" was $56 billion—representing 7 percent of the total gross output value of industry in China and second only to "electronics and telecommunications equipment." China is among the world's largest producers of some chemical products, such as dyes and pesticides. Nonetheless, China is also the world's largest chemical importer, because China's chemical industry is unable to satisfy domestic demand in certain commodities, particularly petrochemicals. In 1993, for example, China imported half of the plastics consumed domestically, and chemical fertilizers have been among China's top five import commodities. (National Bureau of Statistics 1999, pp. 432, 588, 589–590; Young and Wood 1994, p. s1; Wood and Young 1996, p. s1; *Economist* 1994.)

The largeness of China's chemical market has actually hampered the technological upgrading of China's chemical industry, as efforts to protect domestic producers have blocked foreign investment in certain areas. Poor protection of intellectual property rights has also discouraged foreign investors. In the past few years, however, the government has been removing barriers to foreign investment in the chemical industry. As a result, a number of world-scale joint ventures are now being planned, including a $4 billion petrochemical complex by BASF and a purified terephthalic acid project by Amoco. Overall, China expected to have about $10 billion in foreign investment in the chemical industry by the end of 2000. (*Economist* 1994; Rotman 1995, p. s10; Wood and Young 1996, p. s1; Young and Wood 1994, p. s1; *China Business Review* 1995, p. 29.)

China's inability to expand domestic capacity rapidly enough to meet demand may indicate that the chemical industry is experiencing technical difficulties in constructing facilities to produce certain kinds of chemicals, but much of the problem also appears to be simply financial. And whatever limitations China is suffering in its ability to expand production capacity, its basic R&D capabilities are viewed as quite strong. Indeed, a number of Western companies, including DuPont, Hoffmann-La Roche, and Phillips Petroleum, have been licensing technologies developed in Chinese laboratories. China is considered to be particularly strong in basic research in areas such as physical chemistry and polymer science, and Chinese researchers have been characterized as "extremely prolific in new molecule discovery." (Wood and Young 1996, p. s1; *Economist* 1994; Rotman 1995, p. s10; Rotman 1994, p. s26.)

China is considered weaker in experimental chemistry and in developing production processes, although there have been some advances in these areas as well. In general, as is true in many technology areas in China, there is an inability to turn research results into commercial products. And despite China's strength in some areas, senior chemical officials in China acknowledge that, overall, China's R&D lags that of the industrialized countries by about 10 years. (Rotman 1994, p. s26; Cheng 1990, p. 39; Rotman 1995, p. s10; *Chemical Week* 1993, p. s27; Roberts 1996b, p. 39.)

AVIATION

Aircraft, including combat aircraft, transports, and helicopters, are an essential component of a modern military. Most of the technologies used in the design, integration, and manufacture of commercial aircraft are also used for military aircraft. Moreover, many components and technologies used in manufacturing aircraft are also employed in other military systems, such as cruise missiles. Sophisticated civilian aviation technology, therefore, could contribute to efforts to develop a number of important military systems. (OUSD(A&T) 1996, p. 1-1.)

There are roughly a dozen aircraft manufacturers in China. In addition to combat aircraft and jet trainers, these companies produce transports, light and ultralight aircraft, agricultural aircraft, and helicopters. The combat aircraft are mostly based on 1950s and 1960s

Soviet technology, however, and other than co-assembly of McDonnell Douglas passenger aircraft (see below), China's capability for producing transport aircraft is limited to short-range and medium-range turboprops. Chinese helicopters in production are all based on European models. (Jackson 1998, pp. 52–76.)

China's government is aggressively trying to upgrade its civilian aircraft industry. Since the failure of an effort to reverse-engineer a Boeing 707 in the 1970s, these upgrading efforts have focused on cooperation with foreign producers. Using access to its potentially huge aviation market[12] as a bargaining chip, China has succeeded in getting foreign aircraft manufacturers to involve China in the design and production of commercial airliners. The first such agreement involved co-production of the McDonnell Douglas MD-82 in Shanghai in 1985. Between 1985 and 1994, Shanghai Aviation Industrial Corporation (SAIC) assembled 35 MD-82 and MD-83 jetliners; most of these were sold to customers in China, but five were sold to TWA in the United States. To ensure that all of these aircraft would receive U.S. Federal Aviation Administration certification, McDonnell Douglas completely renovated SAIC's factories, provided huge amounts of technical data, and had U.S.-based McDonnell Douglas employees provide 55,000 man-hours of technical training in engineering, tooling, and other areas. Chinese factories provided an increasing proportion of the components used to construct the aircraft, going from 15 percent at the beginning of the program to 50 to 60 percent at the end. Under the canceled MD-90 Trunkliner program, Chinese manufacturers were to have produced 70 percent of the aircraft's parts—essentially everything except the engines and avionics—by the time the last aircraft rolled off the assembly line in 2000. Chinese manufacturers have also been supplying components for a number of other aircraft companies, including tail sections, doors, and fin fairings for various Boeing and Airbus aircraft. (Lewis 1996, p. 32; Mecham 1993, p. 29; Mecham 1995a, p. 27; Mecham 1995b, p. 56; Kahn 1996, p. 1; Jackson 1998, pp. 52–76; Wang 1995, p. 17; Dorminey 1998, p. 81; Mecham 1995c.)

---

[12]One Airbus study forecast that China would need 1000 new airliners worth $100 billion between 1995 and 2014, and some observers have predicted that China will represent 10 percent of the world market for airliners by the end of that period (Macrae 1996, pp. 26–27).

China is also benefiting from Western technology in aircraft engines and avionics. One noteworthy joint venture involves the manufacture of components for Rolls-Royce engines in Xian. This facility began producing advanced turbine blades in 1998, with plans calling for it to be responsible for complete engine modules by 2000. As part of the arrangement, Chinese engineers have been working on secondment at BMW Rolls-Royce's Dahlewitz factory in Germany. The Chinese partner in this joint venture, Xian Aero-Engine (XAE), had already acquired sophisticated lathes, milling and broaching machines, and testing equipment as part of a co-production arrangement with Rolls-Royce to produce Spey engines in the late 1970s. XAE subsequently continued to acquire imported equipment and, in the 1990s, produced components for Rolls-Royce, General Electric, Pratt & Whitney, and other Western engine manufacturers. (Xinhua 1997; *Aviation Week and Space Technology* 1996, p. 61; Paloczi-Horvath 1997; Xinhua 1998; Bailey 1992, p. 30.)

Pratt & Whitney has established several joint ventures to produce aircraft engine components, including a company that manufactures precision sheet metal and other components in Chengdu, a joint venture with XAE to produce compressor airfoils and precision components for commercial aircraft engines and industrial gas turbines, and a company that produces gas turbine engine components in Changsha. In addition, several Western companies are producing avionics systems and components in China, including transponders, air data computers, and weather radar subassemblies. (Bangsberg 1998, p. 4A; Korski 1998, p. 4; Pratt & Whitney 1997; Pratt & Whitney 1998; Dornheim 1998, p. 72.)

Western aerospace companies engaged in joint ventures with China are said to jealously guard those technologies that provide them with their critical competitive edge, but China's aircraft manufacturers are undoubtedly benefiting from the transfer of technologies short of this level. In addition, Chinese companies have been acquiring advanced machinery and tooling equipment and are building state-of-the-art research facilities. As a result, while Chinese observers lament how far behind the West the Chinese aerospace industry is, some Western observers state that China has been developing "with precocious speed." (Berent 1994, p. 102; Paloczi-Horvath 1997; Covault 1996a, p. 32; Todd 1995, p. 18.)

It is also noteworthy (and probably not a coincidence) that many of China's joint ventures with foreign manufacturers, which are for civilian products, involve China's military aircraft producers. For example, the primary subcontractors for the MD-82/83 were the three companies producing or developing China's most advanced combat aircraft, rather than, say, the Shaanxi Aircraft Company, which produces China's largest indigenous transport aircraft, the Yun-8 turboprop.[13]   And many Chinese engineers were transferred to military projects after being trained at SAIC by McDonnell Douglas. Similarly, XAE, joint venture partner with both Rolls-Royce and Pratt & Whitney, produces engines for combat aircraft, and China National South Aero-Engine Company, Pratt & Whitney's joint venture partner in Changsha, is the manufacturer of the WP-11 engine used in Chinese cruise missiles. In some instances, even more overt efforts to exploit foreign aviation technology for military purposes have occurred. The best known example consists of the diversion to the Nanchang Aircraft Company, a producer of cruise missiles and combat aircraft, of machine tools intended for McDonnell Douglas' joint venture in Shanghai. (Mecham 1993, p. 29; Jackson 1998, pp. 52–76; General Accounting Office 1996a; Lachica 1995; Holloway 1997, pp. 14–16; Kahn 1996, p. 1.)

SPACE

Civilian space technology has direct applicability to military systems. In fact, 95 percent of space technologies are said to be dual-use. Areas of potential application include not only military satellites and ballistic missiles, but also more distantly related systems such as aircraft, cruise missiles, and precision-guided munitions, which overlap in such technology areas as propulsion, aerodynamics, and guidance. (OUSD(A&T) 1996, p. 17-1.)

China has an impressive launcher capability for a developing country. In 1970, it became the fifth country (after the Soviet Union, the

---

[13]Although the Chinese government may deliberately guide foreign joint venture partners to producers of defense articles, in many cases these firms are also the preferred choice from the perspective of the foreign investor, as they have historically enjoyed priority in receiving state investment and human resources (personal communication from Richard P. Suttmeier, University of Oregon; and see Huchet 1997, p. 282, for a similar comment with regard to the electronics industry).

United States, France, and Japan) to acquire the ability to launch satellites. In 1984, it became the third (after the United States and the European Space Agency) to employ a cryogenic (liquid oxygen/ liquid hydrogen) upper stage. The 1984 mission also demonstrated China's ability to deploy multiple satellites from a single launch vehicle. More recently, China received attention for offering commercial launch services to other countries. (Clark 1997b, pp. 8–13, 208–214.)

The lift capacity of China's most powerful launchers is comparable to that of Russia's Proton or Europe's Ariane 4. Chinese launchers are capable of placing over nine tons into low earth orbit (LEO) and over five tons into geosynchronous transfer orbit (GTO). The Chang Zheng (Long March) 3B is capable of placing up to 12 tons into low earth orbit (LEO) and over 4 tons into geosynchronous transfer orbit (GTO). A new launch vehicle is reportedly under development that will be able to place more than 20 tons into LEO. The only application for such a system would be in a manned space program (see below). Chinese launchers are less efficient than those of mature space industries, however, achieving only a third of their payload-to-thrust ratios. Another weakness of China's launcher capability has been a high failure rate. Approximately 20 percent of missions have failed due to launcher problems. (Wang 1996, p. 9; Clark 1997b, pp. 8–13, 208–214; Information Office 2000.)

China's satellite capabilities are less impressive than its launch capabilities. China has developed several types of satellites including communications, photo-reconnaissance, meteorological, imaging, navigation, and experimental satellites. China's first communications satellite, Dong Fang Hong 2 (DFH-2), was launched in 1984. Communications satellite development and production have not been able to keep up with demand, however, and only two indigenously produced communications satellites, both Dong Fang Hong 3 (DFH 3) craft produced in collaboration with Germany's Daimler-Benz, are currently in operation. (A third DFH-3 had to be abandoned because of problems with its attitude control system. Earlier generations in the Dong Fang Hong series have finally burned out after far exceeding their original design lives.) The 24 transponders on these satellites are far short of the estimated 150 transponders China needs, and the shortfall has been partially met by purchasing and launching foreign-built satellites and leasing transponder capacity from consortium-owned satellite providers. In the meantime, more advanced communications satellites are under development in China. (Clark 1997b, pp. 331–332; Zhu 1996, p. 139.)

China also has limited photo-reconnaissance satellite capability. The first craft with a recoverable film module was launched in 1974, although the mission failed because of launcher problems. From 1975 to 1992, however, fourteen consecutive missions resulted in successful recoveries, a record achieved by only two other nations. The lifetime of these satellites, which are used for both civilian and military purposes, has increased from 3 to 15 days. The latest models carry 2000 meters of film and have a resolution capability of 10 meters or less. These satellites also carry microgravity experimental payloads. (Clark 1997b, pp. 11, 411; Covault 1996b, p. 22.)

China has recently developed remote sensing satellites capable of transmitting images of the earth's surface in near-real time. This capability was first demonstrated by one of China's photo-reconnaissance satellites, and, in October 1999, China launched a Landsat-type "earth resources" satellite developed in collaboration with Brazil. This satellite incorporates imaging sensors with a resolution of 20 meters. In September 2000, China launched a second imaging satellite, this one apparently developed entirely domestically (although most likely drawing on the technology developed in collaboration with Brazil). (Xinhua 2000.)

Other satellites developed and launched by China include a small number of meteorological craft and several scientific satellites. In October 2000, China launched the first of what is expected to be a series of navigation satellites similar to those of the U.S. Global Positioning System. (Clark 1997b, pp. 12, 411–412; *South China Morning Post* 2000.)

Since 1992, China has had a little-publicized manned space program that aims to put astronauts in space by 2002, which would make China the third country with this capability. Chinese astronauts have been at the Russian cosmonaut training center since 1996, and in November 1999 China tested an unmanned version of the spacecraft that will eventually carry human astronauts. The vessel, based in part on Russian Soyuz designs, weighs 8.4 tons and can carry at least two people. Space shuttle designs have also been described and may be used in the future. (Covault 1996b, p. 22; Clark 1997b, p. 13; Clark 1997a; Laris 1999, p. A1.)

SUMMARY

China has significant production capabilities in all but one (biotechnology) of the eight major industries examined. However, China also has significant limitations to its capabilities in all eight. China's open economic policies and the globalization of production have given China access to all but the most advanced dual-use technologies, and it assembles many high-technology goods, but in most cases China has yet to capture the critical technologies embodied in these goods. Thus, for example, while Chinese companies are able to assemble microcomputers comparable to those produced by IBM or Compaq, they cannot produce those computers without imported microchips. And while China produces some microchips that are near state of the art, it cannot make those chips without imported lithography tools. Many of the sophisticated consumer electronics and computer parts that China exports are merely assembled in China from imported components. And much of China's advanced imported equipment is underutilized, often because the importing organization is unwilling to pay for imported supplies or maintenance. Finally, a number of sectors suffer from a disjuncture between basic research and commercialization, with the result that even though impressive laboratory results are often reported, production technology remains backward.

None of this is particularly surprising. China is one of the poorer countries in the world, with a per capita income well below that of Mexico or Brazil (World Bank 1997b, p. 129). Current capabilities reflect a significant improvement over the past two decades, however. When China's economic reform program began in the late 1970s, Chinese industrial technology was universally obsolescent. Now, while China is hardly a high-tech powerhouse, some sectors are relatively modern. The key question addressed in the next chapter is, how much more improvement will occur in coming years?

Chapter Four

POTENTIAL FOR FURTHER PROGRESS

The literature on the determinants of technological progress identifies a number of factors that affect a country's ability to acquire or develop new technologies.[1] These may be grouped into four categories: capabilities, effort, incentives, and institutions. *Capabilities* refers to the physical and human capital needed for technological innovation. Physical capital comprises the facilities and equipment needed; human capital comprises the trained personnel involved in technological activities. This training includes formal education, as well as specialized nondegree training programs and knowledge acquired through on-the-job training and experience.

Effort refers to the employment of this capital in activities related to technological improvement. It includes not only formal R&D but also routine production activities, which often lead to significant improvements in efficiency and quality. (On an even more fundamental level, without routine production efforts there will be no imperative to translate laboratory advances into practical applications.) For developing countries, technology licensing and direct foreign investment can replace the domestic technological effort associated with basic R&D (although efforts to assimilate and adapt existing technologies are still required) and are an efficient way to benefit from the results of innovation in other countries.

Capabilities and effort are not enough to ensure technological progress however. There must also be an appropriate *incentive*

---

[1]This chapter draws primarily on Lall 1992 and Felker 1998.

structure, which consists of the macroeconomic environment, competition, and factor markets. The macroeconomic environment includes the rate of growth in the domestic and international economies and that rate's stability; inflation, interest, and exchange rates and their stability; the availability of credit and foreign exchange; and political stability. Instability in any of these factors tends to discourage, due to risk aversion, activities that are inherently risky and generally have payoffs only over the longer term, e.g., technological innovation. For the same reason, high interest rates or restrictions on the availability of credit also discourage technological innovation. In contrast, expectations of high economic growth rates (especially in the domestic economy) tend to encourage technological development, as the future market for the products of innovation is expected to expand.

The second part of the incentive structure, competition, can affect technological progress either positively or negatively. If there is no competition, of course, firms have little incentive to innovate, but if there is competition from more-advanced foreign firms, development beyond technologically simple activities may be deterred. Limited, temporary protection of domestic markets (or subsidization) may be conducive to technological progress. Ideally such protection should be offset by incentives for increased efficiency, such as strong incentives to export and encouragement of domestic competition. For this type of intervention to be effective, however, policymakers must be able to correctly identify those sectors and firms that have the potential for technological advancement, and to end protection or subsidization for those that do not. (Fransman 1986, pp. 75–93.)

As for the third part of the incentive structure, well-functioning, flexible factor markets and correct relative factor prices are also essential to technological progress. Efficient capital markets are needed to ensure that financing is available for long-term investment in risky technology development projects and that capital is priced such that it enables technological innovation but still requires successful results. Efficient labor markets are also needed, so that workers have the incentive to acquire technologically advanced skills and so that firms can hire skilled employees in response to technological opportunities.[2] Finally, efficient technology markets are necessary so that firms can acquire requisite technologies. When market failures occur in these areas, the government may have to intervene to provide financing for technology development or subsidies for worker training. Again, the effectiveness of such policies depends on correctly identifying those technologies or skills for which an unmet demand exists.

The last factor needed for technological progress to occur is properly functioning legal, industrial, training, and technology *institutions.* Legal institutions include those generally needed to support industrial activity, such as ownership laws and laws governing contracts, as well as those of particular importance to technological progress, such as the protection of intellectual property rights. Without the protection of patents and copyrights, individuals and firms are discouraged from engaging in innovative behavior, since the benefits of their research will accrue to others. Industrial institutions include those that promote interfirm linkages, such as business associations, industrial consortia, subcontracting networks, joint research institutes, and strategic technical alliances. Such institutions can sometimes provide the necessary scale and synergies for certain technological undertakings, but their main importance is that they facilitate the knowledge flows needed for technological innovation. Industrial institutions also include government programs that provide support to smaller enterprises, such as technology incubators and high-technology parks, or that help firms to restructure and upgrade. Training institutions include trade schools and non-degree-granting technical training organizations. Technology institutions include the maintenance of national technical standards, technology extension programs, and testing bureaus. (OECD 1997, p. 7; Felker 1998.)

---

[2]Worker mobility is also important because it facilitates knowledge flows between firms (OECD 1997, p. 7).

CAPABILITIES[3]

The physical capital available for technological progress in China is difficult to quantify. Investment in facilities and equipment that contribute to technological advances is not reported separately from investment in other fixed assets (although it may be included in the estimates of overall expenditures on R&D discussed below). Anecdotal information, however, suggests that China's physical capital stock is not optimal for facilitating technological progress. The equipment in most facilities is quite backward, and the few laboratories that have state-of-the-art equipment sometimes find that equipment too advanced to support. (Mervis 1995, p. 1134; Kinoshita 1995a, p. 1148; Kinoshita 1995b, p. 1138.)

Information on China's human capital base, at least with regard to formal education, is more readily available. One measure is the level of expenditure on education. Government expenditure on education represents about 2.5 percent of GNP in China, which is lower than in other Asian countries. In the mid-1980s, the South Korean and Taiwanese governments spent about 5 percent of GNP on education, and India spent nearly 4 percent (see Figure 4.1).[4] (National Bureau of Statistics 1999, pp. 55, 637; Lall 1992, p. 174.)

More specific information is available from enrollment and graduation rates at various levels. Primary education has long been one of China's strengths relative to other developing countries. Although enrollment rates have fluctuated over its history, Chinese statistics show that more than 90 percent of all children of elementary school age have been enrolled in school since the 1970s. This rate has increased steadily, reaching 98.9 percent in 1998. Elementary school

---

[3]The assessments in this section and the next three were done for China as a whole, rather than for each of the eight technology areas examined in Chapter Three. This approach was taken in part because data specific to each technology area are not available, but also, and more important, because technology assets are fungible and can be shifted from one area to another. Scientists and engineers trained or working in one field can switch to other fields, and resources devoted to one field can have spill-over effects into other fields. The result is thus a picture of China's overall prospects for technological progress, not estimates of its potential for progress in each of the eight particular technology areas.

[4]The Chinese government increased its attention to education as part of the Ninth Five-Year Plan (1996–2000), however, and has pledged to do more after 2000.

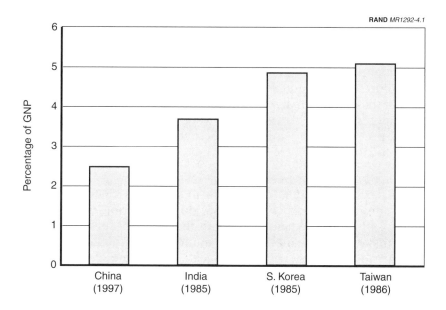

RAND *MR1292-4.1*

Figure 4.1—Public Expenditure on Education as a Percentage of GNP

graduation rates are also high, so at least 79 percent of current 20 to 21 year olds have received a complete primary education.[5] (National Bureau of Statistics 1999, p. 649.)

Secondary education has been more restricted in China, with competitive examinations required for admission and placement. In

---

[5]This percentage was calculated by applying the enrollment rate of 96 percent to the number of children entering elementary school in 1985 and comparing the outcome to the number who graduated six years later, in 1991. The group of children who graduated in 1991 undoubtedly included some who were not in the group that enrolled in 1985—i.e., children who entered elementary school before or after 1985 and graduated ahead of or behind schedule. Conversely, some who did enroll in 1985 undoubtedly also graduated in years other than 1995. Similarly, graduation rates computed on the basis of enrollments six years before also ignore students who enter or leave the population—i.e., emigrate, immigrate, or die. The assumption here is that such effects tend to cancel each other out and that any remaining biases are small. The net effect of such events is most likely to slightly reduce the cohort size six years later, meaning that actual graduation rates are probably slightly higher than those calculated here. These principles also apply to other graduation rates calculated in this report.

1991, only 74 percent of elementary school graduates in China entered middle school. Of these, 82 percent graduated, so about 48 percent of current 20 to 21 year olds have received a middle school education. Of these, only 47 percent went on to enroll in high school, including technical and vocational high schools. Eighty-eight percent of these graduated, so about 20 percent of current 20 to 21 year olds have received a high school education. (State Statistical Bureau 1996, pp. 632, 637; Ministry of Science and Technology 1999, pp. 209–210.)

Higher education has been the most restricted part of China's educational system. In 1980, only 4 percent of all high school graduates were admitted to college. By 1985, this proportion had risen to 22 percent,[6] and it remained at about that level through the 1990s. Nonetheless, this meant that only about 4 percent of current 20 to 21 year olds will receive a college education. Of these, about half will graduate from three-year technical colleges and about half will graduate from comprehensive universities. Chinese leaders, however, recently announced plans to increase college and graduate school enrollment by at least 30 percent. (State Statistical Bureau 1996, pp. 632, 634, 637; Ministry of Science and Technology 1999, pp. 210–212; Lam 2000.)

Only a small fraction of college graduates in China attend graduate school. In 1997, about 6 percent of all college graduates enrolled in graduate school in China.[7] However, a significant number of college graduates have gone abroad for study. According to Chinese statistics, from 1985 to 1993, about 1 percent of all college graduates left to study abroad. Since 1994, this proportion has increased to 3 percent. Only about a third of these students have returned so far, however, so those students who leave to study abroad contribute only partially to China's human capital base. (Ministry of Science and Technology 1999, pp. 212–213; National Bureau of Statistics 1999, p. 644.)

---

[6]More than half of China's middle school graduates in 1985 attended vocational or technical high schools, which are not intended to prepare students for college. If we assume that very few graduates of technical and vocational schools were admitted to college, the admission rate for students attending preparatory high schools was about 47 percent.

[7]This includes graduates both of four-year universities and of three-year technical colleges.

Assuming the most recent enrollment and graduation rates continue, levels of education will improve over time in China. By 2010, about 92 percent of 25 year olds will have received a primary school education, 76 percent will have graduated from middle school,[8] 32 percent will have graduated from high school, and 6 percent will have graduated from college. These relatively high levels (for a developing country) of primary and middle school education will provide China with a good base of literate workers capable of utilizing the relatively simple technologies associated with basic industrialization. As the economy becomes more sophisticated, however, better-educated workers will be needed, both to master more-advanced skills and to facilitate knowledge transfer between engineers and the workforce. In this area, China's human capital base appears to be weak. In Taiwan, for example, about 99 percent of elementary school graduates attend middle school, and 92 percent of middle school graduates enroll in high school. Of high school graduates, 62 percent enter college. China's current elementary and middle school enrollment rates are similar to those of Taiwan in the 1970s, but its high school and college enrollment rates are comparable to those of Taiwan in the 1960s. Thus, China's human capital development appears to be roughly 20 to 30 years behind that of Taiwan. (Council for Economic Planning and Development 1994, p. 268; Lall 1992, pp. 176–177; Directorate-General of Budget, Accounting, and Statistics 1998, p. 85.)

China's greatest weakness is in the area of higher education. Figure 4.2 compares 1990 college education rates in China with those in four other Asian countries that are at various levels of development: Japan, India, South Korea, and Taiwan. Even India, with a per capita income about half that of China (in purchasing power terms) sent more than twice as high a proportion of its college-aged population to college as China did, while Japan, South Korea, and Taiwan sent roughly *fifteen* times as high a proportion.[9] (World Bank

---

[8]The proportion of people graduating from middle school may be even higher, as middle school is now compulsory in China (*Beijing Review* 1996, p. 20).

[9]College enrollment rates in China have improved substantially since 1990 but are still lower than those of any of the other countries considered here. In 1990, 2.1 percent of China's college-aged population was enrolled in institutes of higher education. By 1998, this proportion had increased to 3.2 percent, which is still much less than India's 5.7 percent in 1990. Comparative statistics were not available for years after 1990.

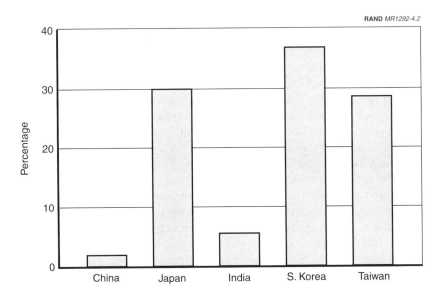

Figure 4.2—Percentage of College-Aged Population Enrolled in Institutions
of Higher Education

1997b, p. 129; State Statistical Bureau 1996, pp. 70, 631; National Science Foundation 1993, p. 62.)

In terms of the absolute number of scientists and engineers produced, however, China, because of its huge population and emphasis on science and engineering in higher education, compares more favorably with other countries.[10] In 1990, China awarded roughly the same number of bachelor's degrees in science and engineering each year as India and the United States do, and awarded about one and one-half times as many as Japan, three times as many as South

---

(National Science Foundation 1993, pp. 61, 95; National Bureau of Statistics 1999, p. 641.)

[10]Some scientists and engineers are needed for routine production, as opposed to development of new technologies, and this number is presumably roughly proportional to the size of the industrial labor force. Thus, the total number of scientists and engineers produced undoubtedly overstates the number of scientists and engineers available for technology development in large countries like China and India, and understates the number available in smaller countries like South Korea and Taiwan.

Korea, and ten times as many as Taiwan (see Figure 4.3).[11]   The comparison was less favorable, however, in advanced (master's and doctoral) degrees. The United States awarded three times as many advanced degrees in science and engineering as China did, while Japan and India awarded nearly twice as many. Nonetheless, China still awarded three times as many advanced degrees as South Korea did and five times as many as Taiwan (see Figure 4.4).

The comparison were still less favorable when only doctoral degrees were considered. The United States awarded nearly ten times as many doctorates in science and engineering as China did, and India awarded nearly three times as many. Japan awarded about the same number as China, however, while South Korea awarded fewer than half as many and Taiwan only about one-seventh as many (see

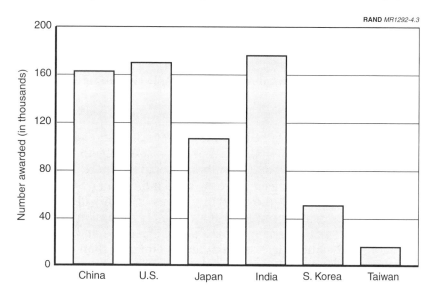

Figure 4.3—Number of Bachelor's Degrees Awarded in Natural Science and Engineering

[11]These figures include mathematics. By 1998, the number of bachelor's degrees in science and engineering awarded in China had increased by 37 percent over the 1990 figure. Data were not available for other countries after 1990. (National Science Foundation 1993, p. 64; National Bureau of Statistics 1999, p. 646.)

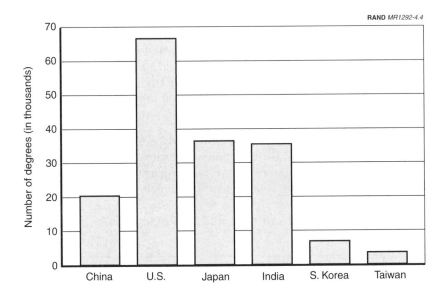

Figure 4.4—**Number of Advanced Degrees Awarded in Natural Science and Engineering**

Figure 4.5).[12] In sum, China is training fewer scientists and engineers than the United States and India are (especially in advanced degrees), roughly the same numbers as Japan (more bachelor's degrees but fewer advanced degrees), and far more than South Korea and Taiwan. Although the scale of China's economy dictates that more of these will be required for routine production activities, China's sheer size means that the pool of scientists and engineers available for technological development is far larger than that which smaller countries such as Taiwan and South Korea can muster. (National Science Foundation 1993, pp. 6, 61–65, 69–71, 75–77.)

---

[12]A large proportion of the graduate degrees in science and engineering awarded in the United States, however, are earned by foreign nationals. More Taiwanese nationals earn science and engineering doctorates in the United States, for example, than in Taiwan. Including doctorates earned in the United States, only about twice as many Chinese as Taiwanese earn doctorates in science and engineering. Many of these people continue working in the United States, however, rather than returning to their home countries. (National Science Foundation 1993, pp. 87–89, 130–132.)

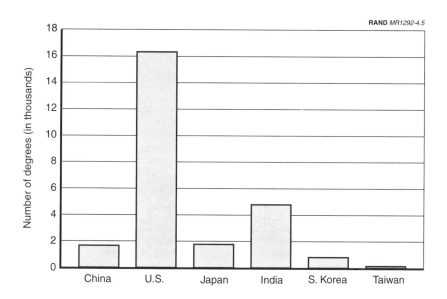

Figure 4.5—Number of Doctoral Degrees Awarded in Natural Science and
Engineering

EFFORT

Technological effort is difficult to measure directly, but indicators include numbers of scientists and engineers in R&D, expenditures on R&D, numbers of scientific and technical publications, and numbers of innovations and patents. Current levels of effort, of course, do not necessarily determine a country's future technological progress, since effort can be increased in the future. Technological progress takes time, however, and current efforts may not yield results for a number of years. China's technological capabilities a decade from now will largely be the result of efforts under way now.

China has a relatively large number of scientists and engineers engaged in R&D—nearly five times as many as India, and seven times as many as South Korea and Taiwan, although only four-fifths as many as Japan and half as many as the United States (see Figure 4.6).

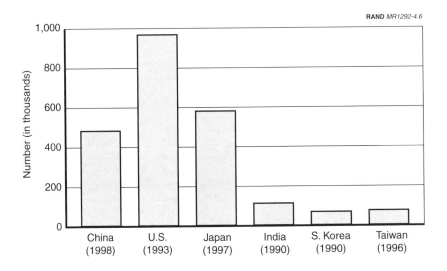

Figure 4.6—Number of Scientists and Engineers in R&D

Relative to the total labor force, of course, the number of scientists and engineers in R&D in China is quite low—less than a fifth as high as in South Korea and Taiwan and less than a tenth as high as in Japan and the United States, although twice as high as in India (see Figure 4.7). (National Science Foundation 1993, pp. 122–123; National Bureau of Statistics and Ministry of Science and Technology 1999, p. 240; Directorate-General of Budget, Accounting, and Statistics, 1998, p. 96.)

Similarly, the absolute magnitude of China's expenditure on R&D (in purchasing power terms[13]) is substantial, almost three times that of South Korea, six times that of India, and nearly ten times that of Taiwan, although only a little over half of Japan's and about one-fourth of the United States' (see Figure 4.8). The comparison is less favorable, however, when considering only R&D funded or per-

---

[13]The purchasing power parity (PPP) factors cited here derive from the "Penn World Tables" (Summers and Heston 1991). World Bank (1997b) estimates of purchasing power parity for China are about a third lower. In both cases, the PPP factors are estimates for the entire economy, not solely the R&D sector. Thus, comparisons of R&D spending between China and other countries should be regarded as approximate.

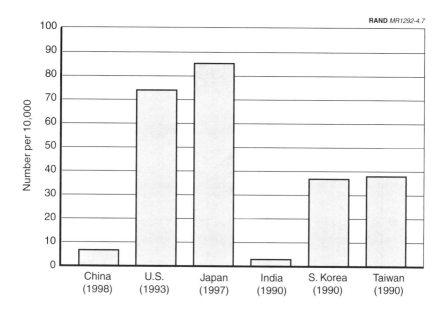

Figure 4.7—Scientists and Engineers in R&D per 10,000 of the
Labor Force

formed by industry (as opposed to by the government, universities, nonprofit research institutes, or foundations). It is generally accepted that R&D performed and financed by productive enterprises is more effective than R&D performed or financed by other organizations. An unusually high proportion of China's R&D is financed and performed by the government (of the countries being compared here, only India has higher proportions). About 60 percent is financed by the government and about 40 percent is performed by government research institutes, compared to less than 20 and 10 percent, respectively, in Japan and South Korea.[14] China's industry-funded expenditures on R&D (in purchasing power terms)

---

[14]Over 40 percent of U.S. R&D is funded by the government, but only 10 percent of it is *performed* by government institutes. Privately financed R&D yields much higher returns than R&D financed by the government does, even when performed by the same enterprises (Griliches 1986).

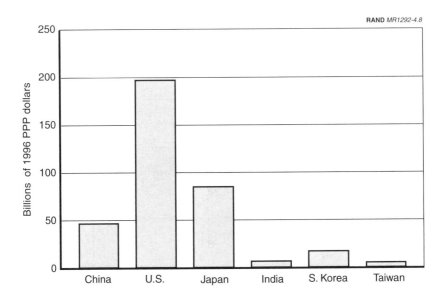

RAND *MR1292-4.8*

Figure 4.8—Total R&D Expenditures

are only a third of Japan's and less than a sixth of the United States', although still one and one-half times South Korea's, more than five times Taiwan's, and roughly ten times India's (see Figure 4.9). In terms of R&D actually performed by industry, China's expenditures (in purchasing power terms) were only about a third of Japan's and a seventh of the United States', although about two-thirds greater than South Korea's, over six times Taiwan's, and more than ten times India's (see Figure 4.10). (National Science Foundation 1993, pp. 96–121; Lall 1992, p. 178; World Bank 1997b, pp. 6–7, 134–135; State Statistical Bureau 1996, p. 661; National Bureau of Statistics 1999, pp. 55, 58, 675; OECD 1999; Directorate-General of Budget, Accounting, and Statistics 1998, pp. 96–97, 151–153; Council of Economic Advisors 2000; National Bureau of Statistics and Ministry of Science and Technology 1999, p. 240.)[15]

---

[15]Figures for all countries are for 1996, except for India's, which are for 1990. Also, data on the amounts of R&D funded by industry were not available for China for years after 1990. Other statistics suggest (Ministry of Science and Technology 1999, p. 171), however, that the *proportion* of R&D funded by industry has not changed much in the

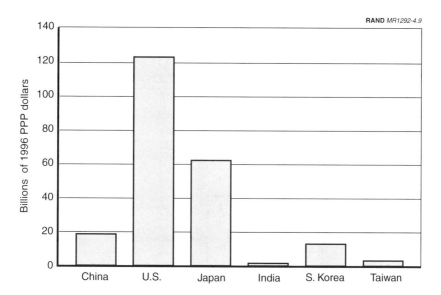

Figure 4.9—Industry-Funded R&D Expenditures

As a proportion of gross domestic product (GDP), China's R&D expenditures are low: Total R&D expenditures in China in 1996 were about 0.6 percent of GDP.[16] Japan, the United States, and South Korea spent nearly 3 percent of GDP on R&D, and Taiwan spent close to 2 percent. Even India spent 0.8 percent (see Figure 4.11). Industry-funded R&D in China compares even more poorly. In 1996, industry-funded R&D represented 0.2 percent of GDP in China, compared to 2.1 percent in Japan, 2.2 percent in South Korea, 1.7 percent in the United States, 1.1 percent in Taiwan, and 0.2 percent in India (see Figure 4.12). Industry-performed R&D also compared poorly, representing 0.3 percent of GDP, compared to 2.0 percent in Japan, South Korea, and the United States, 1.1 percent in Taiwan, and

1990s, so the amounts presented in Figures 4.9 (and 4.12) were estimated assuming the 1990 proportions also obtained in 1996.

[16]The Chinese government called for R&D to be increased to 1.5 percent of GDP by 2000 (Suttmeier 1997, p. 306; Suttmeier and Cao 1999). It is unlikely that this goal was reached, however, as R&D still represented only 0.7 percent of GDP in 1998 (National Bureau of Statistics 1999, p. 675).

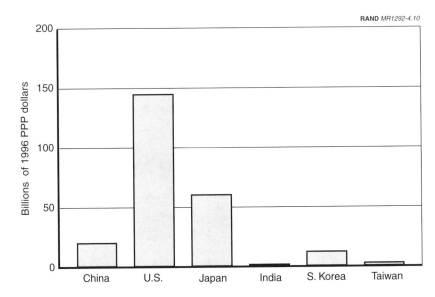

Figure 4.10—Industry-Performed R&D Expenditures

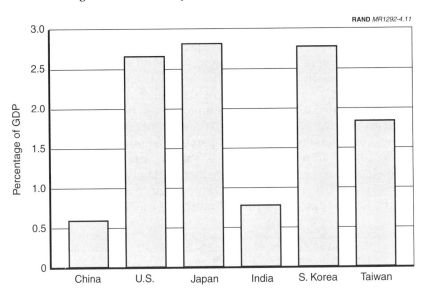

Figure 4.11—R&D as a Percentage of GDP

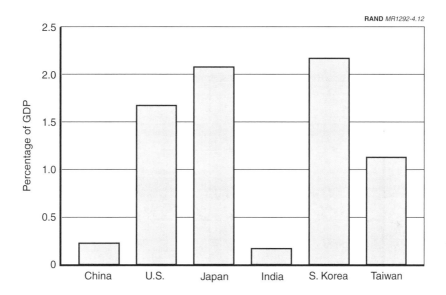

RAND *MR1292-4.12*

Figure 4.12—Industry-Funded R&D as a Percentage of GDP

0.2 percent in India (see Figure 4.13). In these measures, China's R&D spending as a proportion of GDP is again comparable to rates in Taiwan and Korea in the 1970s. (State Statistical Bureau 1996, p. 661; National Bureau of Statistics 1999, p. 675; National Science Foundation 1993, pp. 96–99, 104–107, 112–115, 122–123; OECD 1999; Directorate-General of Budget, Accounting, and Statistics 1998, pp. 96–97, 151–153; Council of Economic Advisors 2000; Lall 1992, p. 178.)[17]

Although China's domestic spending on R&D is low, China enjoys unprecedented amounts of direct foreign investment for a developing country. By bringing in the results of innovation performed elsewhere, direct foreign investment can be a substitute for basic R&D efforts (although efforts to assimilate and diffuse the technology are still necessary) in a developing country. Indeed, as Chapter Three indicates, much of China's advanced technology is due to foreign

---

[17]Figures for India are as of 1990. Industry-funded R&D in China was estimated by assuming that it represented the same proportion of total R&D expenditures in 1996 as in 1990.

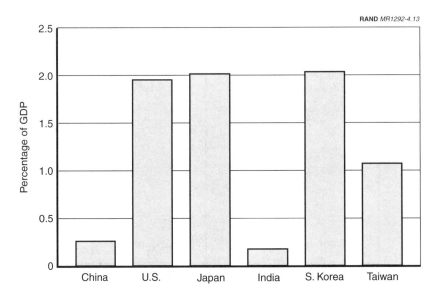

Figure 4.13—Industry-Performed R&D as a Percentage of GDP

investment. The amount of direct foreign investment China has received has been huge in recent years: $45 billion—4.8 percent of nominal GNP—in 1998 alone. By contrast, direct foreign investment in Taiwan has never exceeded 2.3 percent of GNP ($249 million in 1973). Foreign investment is not a direct substitute for domestic R&D, but it is currently an important source of technology for China, and the high levels of foreign investment that China has been receiving (including in relatively high-technology sectors, such as electronics) mean that China's technological progress will be faster than suggested by domestic R&D spending alone.[18] (State Statistical Bureau 1996, pp. 42, 580, 598; Council for Economic Planning and Development 1994, pp. 1, 244.)

Numbers of scientific and technical publications are another measure of the amount (and productivity) of effort being put into technological development (Wagner 1995, pp. 30–40). In 1997, Chinese authors published about one-thirteenth as many articles in interna-

---

[18]Moreover, an increasing amount of foreign investment is now funding R&D in China.

tionally recognized scientific and technical journals as American authors did, and one-third as many as Japanese authors, although they published about one and one-half times as many as Indian authors and twice as many as Taiwanese and Korean authors (see Figure 4.14). (Ministry of Science and Technology 1999, p. 217.)

Because standards and legal systems differ, the absolute number of patents granted is not necessarily an accurate measure of a country's level of technological effort. The proportion of patents awarded to domestic versus foreign applicants is a more revealing measure, with developing countries, ceteris paribus, tending to grant a high proportion of patents to foreign applicants. In 1997, 44 percent of the patents granted in China for "creations and inventions" originated domestically. This was a significant increase over 1990, when only 30 percent of patents granted were domestic in origin. By comparison, 69 percent of South Korean, 56 percent of Taiwanese, and

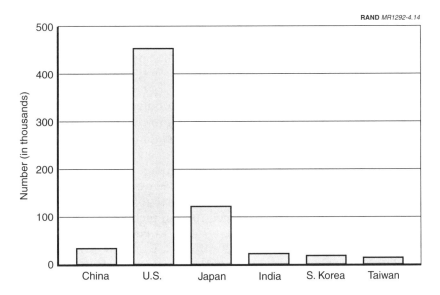

Figure 4.14—Number of Publications in International Scientific and
Technical Journals

20 percent of Indian patents granted were domestic in origin in 1986. (Lall 1992, pp. 174–175, 178–179; Ministry of Science and Technology 1999, p. 196.)

INCENTIVES

The macroeconomic environment, competition, and factor markets provide mixed incentives for technological innovation in China. Some aspects of the *macroeconomic environment* stimulate technological development; while others inhibit it. The high economic growth rates China has been experiencing, if they continue, are a major stimulus, as they mean that the market for new or improved products is rapidly expanding. Growth rates have tended to fluctuate greatly, however, and uncertainty about the domestic economy may discourage innovation. Similarly, inflation, while currently low, has also fluctuated greatly in China, creating uncertainty about the expected real costs of capital, which tends to discourage long-term technology development projects.

China maintains a fixed exchange rate between the renminbi and the dollar. This promotes near-term stability in the exchange rate (although the renminbi is prone to devaluation every few years), enabling enterprises to form reasonable expectations about future prices of imported inputs and exported products, which should be conducive to investing in technological improvements. The availability of foreign exchange and commercial credit, however, have been problems. Access to foreign exchange is generally restricted, so firms must generate their own foreign exchange through export sales if they wish to purchase imported machinery or components. This limits the ability of firms (other than those that are foreign owned or joint venture) to acquire advanced foreign technology, unless they are export-oriented companies. Access to commercial credit for technological development is also limited, primarily because most bank loans are used to keep loss-making state-owned industries solvent. Seventy-six percent of the lending by China's four largest banks—which made 91 percent of all domestic-currency commercial bank loans in China in 1997—is to state-owned enterprises, the least likely place for technological innovation to occur (Lawrence and Saywell 1998, p. 58). Finally, political instability has also discouraged technological innovation in China. Frequently changing policies, as well as outright turmoil such as in 1989, have contributed to a strong short-term orientation in the Chinese economy. Entrepreneurs are looking for investments that provide immediate returns, not technology development projects whose payoffs may be several years in the future.

Competition also presents a mixed picture in China with regard to incentives for technological innovation. Various restrictions on imports protect local industry from foreign competition. Foreigners are not banned outright, however, in most industrial sectors; and for many state-owned enterprises, imported products (or products produced domestically by foreign-owned or joint-venture enterprises) are their only competition (Suttmeier 1997, p. 312). More important, China's industries have strong incentives to export their products (exporting provides access to foreign exchange and investment, etc.), which forces them to be competitive on the international market. Furthermore, China's entry into the World Trade Organization (WTO), expected later this year, may increase the amount of competition from foreign products in China's domestic markets (although the examples of Japan and other countries suggest that being a WTO number does not always mean unrestricted access to a country's domestic markets).

Domestic competition varies by sector. Many sectors are highly competitive, and this has stimulated a strong demand for improved technology (Yuan 1995, p. 227; Suttmeier and Cao 1999). In other sectors, however, a lack of competition due to local protectionism has resulted in a large number of small firms. For example, China had 140 independent television manufacturers in the early 1990s (Huchet 1997, p. 256). Under such circumstances, firms not only lack the scale necessary to support organic R&D, they also lack the incentive to innovate that competition would provide. In the 1990s, the government attempted to mitigate this problem by consolidating state-owned firms into large business conglomerates (*jituan*) modeled on the Japanese *zaibatsu* and Korean *chaebol*. China's state-owned industries, however, which still produce nearly half of total industrial output, suffer from additional problems. Hand in hand with their enjoyment of protected markets and their guaranteed supply of capital, materials, and labor are the absence of a profit motive, limitations on the operational rights of managers, and the absence of a link between worker contributions and compensation.

This all means that they have little incentive to reduce costs, improve quality, or develop new products, which translates into little demand for new technologies. (Baark 1991b, p. 541; Yuan 1995, p. 229; Ding 1995, pp. 244–246; Suttmeier 1997, p. 311.)

Insufficient competition has not only been an issue with regard to potential *consumers* of technology, it has also been a problem with regard to *providers* of technology. Until the 1980s, virtually all technological activity took place in research institutes subordinate to the central ministries or the state-run Chinese Academy of Sciences. These research institutes received their funding through annual budgetary allocations from the central government, which also designated the research projects that were to be conducted. These projects were often unrelated to production needs, and funding levels for research institutes as a whole, as well as financial remuneration for individual researchers within the institutes, were unrelated to their record of technological accomplishment. This was true even of R&D institutes attached to productive enterprises. It has been estimated that, as a result, less than 10 percent of research results in the 1970s were put into production (Yuan 1995, pp. 214–216, 230; Baark 1991b, p. 533).

Since the 1980s, however, the personnel and funding systems of China's state-run research institutes have undergone significant reforms that affect the incentives to produce useful innovations. The tenure system has been abolished, making it possible (in theory) for enterprises to dismiss unproductive researchers. At the same time, a system of prizes for major contributions has been instituted, and efforts are made to ensure that resources flow to productive individuals and organizations. The most important reform, however, has been a significant reduction in the amount of funding provided by budgetary allocations. Research institutes have been forced to rely either on winning publicly bid contracts with the government for important research projects, or, increasingly, on nongovernmental sources of funding. (Baark 1991b, p. 534; Suttmeier 1991, p. 558; Yuan 1995, pp. 218–219, 220, 221; Suttmeier 1997, p. 306; Suttmeier and Cao 1999; Kinoshita 1995c, p. 1143; Zhou 1995, p. 1153.)

Research institutes have responded to the reduction of state allocations in a number of ways. Some contract with commercial enterprises to provide technology transfer, R&D, or technical services.

Others have established independent subsidiaries that market commercial products or services. Some institutes have even transformed themselves completely into commercial enterprises that no longer receive budgetary allocations from the state. Others have merged into pre-existing commercial enterprises, becoming their R&D laboratories. All of these responses reflect the increased demand for technological innovation—as a result of increased competition in the Chinese market—in combination with the increased competitive pressures on the research institutes themselves. (Baark 1991b, p. 544; Yuan 1995, p. 222; Suttmeier 1997, pp. 310, 320; Gu 1999, pp. 29–52; Suttmeier and Cao 1999.)

Factor markets appear to be the worst part of China's incentive structure with regard to technological progress. Capital markets in China are poorly developed. As mentioned above, most bank lending is government directed, and other lending tends to gravitate toward areas, such as real estate development, with potentially high and immediate payoffs. Stock markets are limited in scope and not a significant source of capital for investment in technological improvements. Overall, only 17 percent of capital distribution in China is estimated to be determined by market forces (Lawrence 1998, p. 22). Furthermore, given the various distortions in China's capital markets, capital that is available is unlikely to be priced correctly. Some efforts have been made to compensate for these problems, however. One of the functions of the government's Torch technology development program has been to broker financing of high-technology start-ups and venture capital companies, and a Bank of Science and Technology has been established for the same purpose. In addition, in June 1999, the State Council approved the trial initiation of a Technology Innovation Fund of one billion yuan ($120 million) to provide venture capital for "small and medium-sized technology-based firms." (Suttmeier 1991, p. 554; Yuan 1995, p. 222; Suttmeier and Cao 1999.)

The labor market in China also suffers from inefficiencies, although less so than in the past. State-owned industries are still constrained in their ability to hire and fire employees, which tends to keep labor mobility below the level desirable for technological development. Workers are no longer forcibly assigned to work units, however, and township, collective, private, foreign-owned, and joint-venture enterprises have more flexibility than state-owned enterprises do in hiring and firing employees. As a result, the most talented workers tend to flow to the most dynamic nonstate enterprises. Cities such as Shenzhen are known for their high proportion of college-educated workers who move freely between enterprises. Thus, while the labor market for state-owned industries does not operate efficiently, that for the nonstate sector—the sector most likely to produce technological progress—operates relatively well. Unfortunately, the majority of R&D funding still occurs in the state-owned sector.

Technology markets in China are underdeveloped. Indeed, they were nonexistent prior to 1978. Under the central planning system, research institutes were expected to simply provide the results of their research at no cost to the productive enterprises that needed them. In practice, however, research institutes had little incentive to ensure that the technologies they developed were of use to industry, much less to transfer them once developed, and the state-run industry had little incentive to seek improved product or process technologies. The hierarchical organization of Chinese industry also meant that technologies were generally only available to industries in the same industrial system (*xitong*) as the research institute. One of the initial technology reform efforts after 1978 was an attempt to promote the diffusion (and creation) of technology by allowing the existence of technology markets so that enterprises could purchase technologies from the research institutes that had developed them. To facilitate such transactions, national and local "technology fairs" have been held since 1981. The domestic technology market in China remains small,[19] however, and most transactions consist of technological consultancy or technology services, rather than pure technology transfers. (Baark 1991b, p 534; Yuan 1995, p. 217; Suttmeier 1997, pp. 315, 320; Gu 1999, p. 50; Suttmeier and Cao 1999.)

INSTITUTIONS

Prior to the 1980s, the Chinese system for technological development was based on the supply-oriented Soviet model, in which research

---

[19]The total value of all domestic technology trade in 1997 was $4.2 billion, as compared to $21.4 billion in technology imports and exports (Ministry of Science and Technology 1999, pp. 200–202).

institutes under the central and local governments and the various industrial ministries conducted R&D and provided the results to production units. This institutional structure had several drawbacks. First, because the research units were separated from the production units, there was little connection between research efforts and production needs. Technical problems arising from production were not resolved, and the majority of research findings were not applicable to production. Moreover, knowledge generally flowed only one way—from the research institutes to the production units, and research institutes failed to benefit from the considerable knowledge developed in the course of routine production. Finally, the hierarchical organization of this system meant that opportunities for horizontal knowledge exchange between research institutes or industrial sectors were limited. Technological progress is as much the result of fortuitous convergences of knowledge as deliberate research efforts, so the lack of horizontal interchanges between researchers attempting to develop new technologies was a major failing of this type of organization. The resulting system was capable of producing incremental improvements in the areas on which it focused, but was unlikely to spontaneously produce qualitative technological advances.[20] (Baark 1991b, pp. 532–533, 537; Yuan 1995, pp. 214–216; Suttmeier 1997, pp. 315, 320; Gu 1999, pp. 9–11.)

Although this system partially persists in China (roughly 60 percent of R&D is still funded by the government), the reforms that occurred beginning in the 1980s have done much to improve the institutional context for technological development. In particular, reducing budgetary allocations to research institutes, which has caused many of them to turn to commercial markets for their revenues, has fundamentally altered the relationship between those institutes (or the subsidiaries they have spun off) and industrial producers. Such entities now interact directly with their clients, with the result that research is tailored to client needs and knowledge actually can flow from production enterprises to the research institutes instead of only from the research institutes to the production enterprise. Moreover,

---

[20]See Cliff 1997 for a study of technological development in China's steel industry. Qualitative advances did, of course, occur in technologies—primarily weapons technologies—that the government made priority projects and supported with large amounts of resources.

research institutes are free to find clients in any industrial sector, so knowledge can be exchanged between research units and across industrial sectors. Indeed, many research institutes have transformed themselves into (or spun off) enterprises that, rather than performing traditional R&D, provide technical consulting, engineering services, and technology brokering, filling important institutional niches that were vacant prior to the 1980s. (Baark 1991, p. 542; Yuan 1995, p. 222; Suttmeier 1997, pp. 310, 320; National Science Foundation 1993, pp. 98, 102.)

In addition, the Chinese government has undertaken a number of initiatives that strengthen the institutional environment for technological progress in China. One of these has been to establish a number of "technology enterprise service centers" that provide technology start-ups with services and information in areas such as finance, equipment procurement, marketing, and tax regulations. The government also has provided management training for high-technology firms and established a number of "science parks"—technology development zones in which high-technology firms are provided with superior infrastructure support and incentives such as tax holidays. This not only encourages the establishment of high-technology enterprises, but, because many such enterprises are closely located to each other, also promotes the technological cross-fertilization that China has traditionally lacked. (Suttmeier 1991, p. 554; Baark 1991b, p. 544; Conroy 1992, pp. 10–17; Qin 1992, p. 1128; Suttmeier 1997, p. 317; Suttmeier and Cao 1999.)

Other institutions supporting technological development in China remain weak. A patent law was promulgated for the first time in 1985, but the legal system remains problematic, with contracts frequently unenforceable and violations of intellectual property rights common. The unenforceability of contracts inhibits the efficient functioning of the market, especially for intangible goods such as technology, while rampant violations of intellectual property rights discourage investment in innovation. Industrial institutions are also weak, with business associations and other mechanisms for interfirm linkage remaining underdeveloped. Such institutions are growing, however, and should continue to grow in coming years.

SUMMARY

China's prospects for technological progress are mixed. Its physical and human capabilities are substantial but uneven. Facilities and equipment tend to be either backward and poorly maintained or more advanced than can be effectively used. China has a solid primary education base for a developing country, but secondary and higher education rates compare poorly with those of more-developed countries such as South Korea and Taiwan, much less with highly developed countries such as the United States and Japan. The huge size of China's population means that China is training scientists and engineers in numbers comparable to those of the United States and Japan, which gives China a much larger human capital base for R&D efforts than South Korea and Taiwan have.[21] But the equally huge size of China's workforce means that a greater proportion of scientists and engineers are required for routine production activities rather than being available for technology development. Thus, China's total human capital resources still compare unfavorably with those of the United States and Japan.

Technological effort is also limited in China. The total number of scientists and engineers working in R&D is smaller in China than in the United States and Japan, although far greater than in India, Taiwan, and South Korea. Moreover, as a proportion of the total labor force, the number of scientists and engineers in R&D is much lower in China than in any of these countries except India. Similarly, spending on R&D is much lower in China than in the United States and Japan although much greater than in India, Taiwan, and South Korea. With regard to scientific and technical publications, China is well behind the United States and Japan, though well ahead of India, Taiwan, and South Korea.

The incentive environment in China is also less than optimal for technological progress. Fluctuating inflation rates, limited access to credit and foreign exchange, and periodic instability all tend to discourage investment in new technologies. Chinese firms are encouraged to compete on the international market, but protected domestic markets reduce the incentive for technological innovation.

---

[21]India is also producing scientists and engineers in numbers comparable to the United States and Japan.

Underdeveloped capital, labor, and technology markets also inhibit technological progress. The institutional structure needed for technological progress—such as the protection of intellectual property rights and linkages between R&D institutes and productive enterprises—is improving, but at this point it, too, remains underdeveloped.

The net implication of these factors is that China can expect to make significant technological progress in coming years but cannot possibly catch up to, much less "leapfrog," the United States or Japan in the foreseeable future. By many measures, China's prospects for technological progress appear comparable to those of Taiwan and South Korea in the 1970s, particularly when comparing education levels and spending on R&D, although China's incentive and institutional structures are probably somewhat worse. If China's economy and technology develop over the next 20 years at a rate similar to those sustained by Taiwan and South Korea during the 1970s and 1980s, average technological levels in China by 2020 might be roughly comparable to those in Taiwan and South Korea in the 1990s. There could be a difference, however, because of China's scale. Average technological levels in South Korea and Taiwan still lag well behind those of Japan and the United States. South Korea and Taiwan are, however, technological leaders in specific niches. If China follows a similar developmental path, its huge size (China's population is 25 times that of South Korea and 50 times that of Taiwan) suggests that the number of such niches would be much larger. Thus, while China will on average still be technologically well behind the United States and Japan in 2020, it could have state-of-the-art technological capabilities in a substantial number of areas by that time.

Much depends, of course, on the strategy for economic and technological development that China pursues in coming years. The rapid economic and technological progress that Taiwan and South Korea have achieved is primarily a product of their open economies, which have enabled them to acquire and absorb technologies largely developed elsewhere. For small countries like South Korea and Taiwan, this is the only viable choice. A large country such as China, however, has the option of relying to a greater degree on indigenous technology development. Although China is unlikely to return to the technological autarky of the 1960s and 1970s, national pride and concerns about national security—along with Western technology controls—could lead China to rely more on indigenous efforts, at least in key sectors deemed vital to China's national interests. This would likely lead to China being less dependent on foreign technology inputs, but having overall technological capabilities that are less advanced than they would be otherwise. Beijing's recent efforts to be admitted to the WTO suggest that China's leaders are choosing the path of openness and integration (without, of course, forgoing state-led efforts in key technology areas). Even so, China's ongoing problems of economic reform, particularly the continuing burden of unproductive state-owned industries, mean that China will be fortunate if it is able to emulate the technological success Taiwan and Korea have seen over the past 20 years.

Chapter Five

CONCLUSION

China's technological capabilities have increased dramatically since its reform program began in the late 1970s. At that time, China's industrial technology was largely based on 1950s-era Soviet technology and thus was 20 or more years behind that of the rest of the world. The gap has narrowed considerably since then, with a few areas now at or near state of the art. According to one estimate, technological progress was about four times as rapid in China as in the advanced industrial countries during the 1980s (Wu 1995, p. 219), and this pace undoubtedly continued or even accelerated during the 1990s. Nonetheless, critical gaps exist in China's technological capabilities in areas such as lithography tools for integrated circuits (ICs), jet engines, and nuclear reactors. Overall, China's commercial technology remains well behind that of the United States and other advanced countries. This gap will close further over the next two decades, but average technological levels in China will remain significantly behind those in the United States and Japan even by 2020.

Many of the areas in which China is acquiring advanced foreign technology—such as electronics, nuclear power, and aviation—have military applications. Moreover, in 1986 the Chinese government embarked on an ambitious High-Technology Research and Development program whose aim was to achieve world-class capabilities in seven technology areas that, while not explicitly military, have obvious military implications.[1] In coming years, the combina-

---

[1] The seven areas were biotechnology, space, information technology, lasers, automation, energy, and advanced materials (Humble 1992, p. 7).

tion of foreign technology transfer and domestic research efforts could begin to make technologies available to China's defense industries that are very similar to those available to U.S. defense industries in some areas.

Nonetheless, China's overall military technology in 2020 will still be significantly inferior to that of the United States, for several reasons. First, as just noted, China's average level of commercial technology will still lag behind advanced world practice. Second, because development cycles for weapons are long, military systems are often designed around technologies that are a decade or more old by the time the weapons become operational.[2] Thus, the military systems that the United States and China field in 2020 will largely reflect the technologies available to those countries in 2010 or earlier. Finally, the process of translating civilian technological capabilities to military technology is nontrivial. Even though military systems build on technologies that are fundamentally civilian, they still involve technologies that are specifically military and thus must be independently developed. Furthermore, even if all the component technologies of a weapon system are available, the process of integrating them into a smoothly functional whole is challenging. This has been demonstrated, for example, by the difficulties Japan's defense industries have experienced in developing the F-2 indigenous fighter aircraft. (Lorell 1995.)

Simply because China's technological capabilities will lag those of the United States does not mean, however, that China could not present a serious military challenge to the United States. The Soviet Union's relative technological backwardness did not prevent it from deploying a military that looked extremely threatening to the West during the Cold War. Part of the credit must go to the sheer quantity of forces involved, but part must also go to the fact that Soviet military systems, although generally inferior to their Western counterparts, nonetheless compared far more favorably than the overall technological levels of the two societies suggested. This resulted from the excellence of the Soviet design bureaus and the relative efficiency of the Soviet weapons acquisition process. It also resulted,

---

[2]In the United States, 13 to 15 years typically elapse between the initiation of a major weapon development program and the initial operational capability of the first production units.

however, from the substantial amounts of resources the Soviets devoted to military R&D—as much as 2 to 3 percent of GNP. For China, this level of military R&D spending would roughly double the size of its military expenditure and would impose a considerable burden on the economy as a whole and on government budgets in particular. China's leadership does not appear likely to pursue such a course at present. (Holloway 1983, pp. 114, 118, 132–140; IISS 1995.)

As an alternative to the Soviet model of competing with the United States across the board in military capabilities, China could develop "niche" capabilities that would present difficulties for the U.S. military in specific scenarios because of the asymmetrical conditions under which the Chinese and U.S. militaries would operate. For example, if the Chinese were to attack Taiwan and the United States were to intervene, China would have the advantage of operating out of bases on mainland China. The United States would have to operate far from home, its forces constrained to basing on Taiwan itself, on a small number of nearby islands, or at sea—all of which would limit the numbers and types of military systems the United States could employ. In addition, political considerations might limit the types of missions U.S. forces could engage in against China. With such restrictions in play, having a few specific types of systems with strong capabilities could provide China with a military advantage.

To summarize, then, the Chinese military will not be the technological equal of the U.S. military by 2020. Nonetheless, the U.S. military, including the U.S. Air Force, must prepare for a Chinese military whose technological capabilities will steadily continue to advance. This steady advance means that China will increasingly be able to produce systems that are recognizably modern and that Chinese capabilities may approach or equal those of the United States in some areas. While there is no need for alarm about China's technological potential in the next 20 years, the United States must prepare for the possibility of conflict with a Chinese military that will become increasingly sophisticated, and must continue to closely monitor China's R&D efforts in order to detect and respond to any programs that represent particular threats to the U.S. ability to carry out combat missions in East Asia.

Anthony, Ian, "Appendix 12A: Transfers of Digital Communications System Technology," in *SIPRI Yearbook 1996: Armaments, Disarmament, and International Security*, Oxford: Oxford University Press, 1996, pp. 552–559.

Arnett, Eric, "Military Technology: The Case of China," *SIPRI Yearbook 1994*, Oxford: Oxford University Press, 1994, pp. 359–386.

Aviation Week and Space Technology, "Rolls Signs with AVIC," 27 May 1996, p. 61.

Baark, Erik, "Computer Software and Biotechnology in the PRC: Analytical Perspective on High-Tech Politics," *Issues and Studies*, September 1991a, pp. 70–93.

Baark, Erik, "Fragmented Innovation: China's Science and Technology Policy Reforms in Retrospect," in Joint Economic Committee, ed., *China's Economic Dilemmas in the 1990s: The Problems of Reforms, Modernization, and Interdependence*, Washington DC: U.S. Government Printing Office, 1991b, pp. 531–545.

Bailey, John, "Engines of Change," *Flight International*, 15–21 January 1992, pp. 29–31.

Bangsberg, P. T., "Pratt Taps China Company for Aircraft Engine Parts," *Journal of Commerce*, 5 March 1998, p. 4A.

Beijing Review, "Outline of the Ninth Five-Year Plan for National Economic and Social Development and the Long-Term Targets Through the Year 2010," 29 April–5 May 1996, pp. 18–21.

Berent, Mark, "Asian Aeronautics," *Asian Defence Journal*, September 1994, p. 102.

BioIndustry Association, *Biotechnology in China: Report on a Biotechnology Fact-Finding Mission to China*, London, 1996.

Brodie, Jonathan, "China Moves to Buy More Russian Aircraft, Warships, and Submarines," *Jane's Defense Weekly*, 22 December 1999, p. 13.

Chemical Week, "China's Research Develops," 25 August/1 September 1993, p. s27.

Chen Junliang, Yan Liemin, and Li Yonglin, "Switching Systems and Switching Software Development in China," *IEEE Communications Magazine*, Vol. 31, No. 7 (July), 1993, pp. 56–60.

Chen Yunqian, "Driving Forces Behind China's Explosive Telecommunications Growth," *IEEE Communications Magazine*, Vol. 31, No. 7 (July), 1993, pp. 20–22.

Cheng, Siwei, "Focusing on R&D in China," *Chemical Engineering*, February 1990, pp. 35–39.

China Business Review, "Chemical Minister Lays out Industry Priorities," July–August 1995, p. 29.

China Daily, "Lucent Expands Telecom Market," 18 August 1997.

Clark, Phillip, "Chinese Designs on the Race for Space," *Jane's Intelligence Review*, Vol. 9, No. 4 (April), 1997a.

Clark, Phillip, ed., *Jane's Space Directory 1997–98*, Coulsdon, Surrey, UK: Jane's Information Group, 1997b.

Cliff, Roger, "Technical Progress and the Development of China's Steel Industry," *The American Asian Review*, Vol. XV, No. 3, Fall 1997, pp. 153–189.

Cohen, Eliot A., "A Revolution in Warfare," *Foreign Affairs*, Vol. 75, No. 2 (March/April), 1996, pp. 37–54.

Cohen, William S., *Annual Report to the President and the Congress*, Washington D.C.: U.S. Government Printing Office, 2000.

Conroy, Richard, *Technological Change in China*, Paris: Development Centre of the Organisation for Economic Co-operation and Development, 1992.

Council for Economic Planning and Development, *Taiwan Statistical Data Book 1994*, Taipei, 1994.

Council of Economic Advisors, *Economic Report of the President 2000* (http://www.access.gpo.gov/eop/, accessed 22 May 2000).

Covault, Craig, "China Accelerating Aerospace Development," *Aviation Week and Space Technology*, 28 October 1996a, p. 32.

Covault, Craig, "Chinese Manned Flight Set for 1999 Liftoff," *Aviation Week and Space Technology*, 21 October 1996b.

Denton, Douglas, "The Wireless Revolution," *The China Business Review*, March–April 1996, pp. 18–21.

Ding Jingping, "Technical Transformation and Renovation in PRC Industry," in Denis Fred Simon, ed., *The Emerging Technological Trajectory of the Pacific Rim*, Armonk, NY: M. E. Sharpe, 1995, pp. 239–255.

Directorate-General of Budget, Accounting, and Statistics, Executive Yuan, Republic of China, *Statistical Yearbook of the Republic of China*, Taipei, 1998.

Dorminey, Bruce, "Chinese Manufacturing Looks for a Second Act," *Aviation Week and Space Technology*, 9 November 1998, p. 81.

Dornheim, Michael A., "Avionics Joint Ventures on the Rise in China," *Aviation Week and Space Technology*, 23 February 1998, pp. 70, 72, 77.

Economist, "On the Back Bunsen Burner," 24 September 1994.

Elegant, Simon, "Comeback Kid," *Far Eastern Economic Review*, 3 September 1998, pp. 10–14.

Erkanat, Judy, and Chad Fasca, "China Gearing Up . . . But," *Electronic News*, Vol. 43, No. 2176, 14 July 1997, pp. 1, 37, 48, 52, 142.

Felker, Greg, "Malaysia's Industrial Technology Development: Firms, Policies, and Political Economy," in K. S. Jomo, Greg Felker, Rajah Rasiah, eds., *Industrial Technology Development in Malaysia: Industry and Firm Studies*, New York: Routledge, 1998.

Fransman, Martin, *Technology and Economic Development*, Boulder, CO: Westview Press, 1986.

Futures, "Biotechnology in China and the USA," April 1989, pp. 227–228.

Gan, Renchu, and Jianjun Zhang, "Computer-Aided Production Management in China," *Computers in Industry*, No. 19, 1992, pp. 113–117.

General Accounting Office, *Export Controls: Sensitive Machine Tool Exports to China* (GAO/NSIAD-97-4), November 1996a.

General Accounting Office, *Export Controls: Sale of Telecommunications Equipment to China* (GAO/NSIAD-97-5), November 1996b.

Geppert, Linda, "Chip-Making in China," *IEEE Spectrum*, December 1995, pp. 36–45.

Granitsas, Alkman, "Local Hero," *Far Eastern Economic Review*, 3 September 1998, pp. 12–13.

Griliches, Z., "Productivity, R&D and Basic Research at the Firm Level in the 1970s," *American Economic Review*, Vol. 76, No. 1, 1986, pp. 141–154.

Gu, Shulin, *China's Industrial Technology: Market Reform and Organization Change*, London: Routledge, 1999.

Hao Weimin, "Public Switching Deployment in China," *IEEE Communications Magazine*, Vol. 31, No. 7 (July), 1993, pp. 52–55.

Hibbs, Mark, "U.S.-China Deal Rumored Reason Westinghouse Staying a U.S. Firm," *Nucleonics Week*, 2 April 1998a, pp. 2–3.

Hibbs, Mark, "Reactor Orders May Be Placed in 1999 for Shandong, Yangjiang," *Nucleonics Week*, 9 April 1998b, pp. 1, 12–14.

Hibbs, Mark, "China's Equipment Makers Aim to Get Pieces of Nuclear Deals," *Nucleonics Week*, 9 April 1998c, pp. 14–15.

Hilborn, Cathy, "Media Darling," *Far Eastern Economic Review*, 30 July 1998, pp. 50–51.

Holloway, David, *The Soviet Union and the Arms Race*, New Haven: Yale University Press, 1983.

Holloway, Nigel, "Cruise Control," *Far Eastern Economic Review*, 14 August 1997, pp. 14–16.

Hu Qili, "China's Electronic Information Industry Faces the 21st Century," *Keji Ribao (Science and Technology Daily)*, 11 September 1997, p. 1, in Foreign Broadcast Information Service, Daily Report: China (FBIS-CHI-97-302), 29 October 1997.

Huchet, Jean François, "The China Circle and Technological Development in the Chinese Electronics Industry," in Barry Naughton, ed., *The China Circle: Economics and Technology in the PRC, Taiwan, and Hong Kong*, Washington D.C.: Brookings Institution Press, 1997, pp. 254–285.

Hui, Saiman, and Hilary B. McKown, "China Computes," *The China Business Review*, September–October 1993, pp. 14–20.

Humble, Ronald D., "Science and Technology and China's Defence Industrial Base," *Jane's Intelligence Review*, January 1992, pp. 3–11.

Information Office of the State Council of the People's Republic of China, "China's Space Activities," Beijing: November 2000.

International Institute for Strategic Studies (IISS), "China's Military Expenditure," *The Military Balance 1995/96*, London: Oxford University Press, 1995, pp. 270–275.

Jackson, Paul, ed., *Jane's All the World's Aircraft*, London: Jane's Information Group, Ltd., 1998.

Johnson, Stuart E., and Martin C. Libicki, *Dominant Battlespace Knowledge: The Winning Edge*, Washington D.C.: National Defense University Press, 1995.

Kahn, Joseph, "Clipped Wings: McDonnell Douglas's High Hopes for China Never Really Soared," *The Wall Street Journal*, 22 May 1996, p. 1.

Kinoshita, June, "Agriculture Finds a Niche: Drug Researchers Seek Help," *Science*, Vol. 270, 17 November 1995a, pp. 1147–1149.

Kinoshita, June, "Government Focuses Funds, and Hopes, on Elite Teams," *Science*, Vol. 270, 17 November 1995b, pp. 1137–1139.

Kinoshita, June, "Incentives Help Researchers Resist Lure of Commerce," *Science*, Vol. 270, 17 November 1995c, pp. 1142–1143.

Korski, Tom, "Pratt and Whitney in US$ 27m Hunan Deal," *South China Morning Post*, 3 March 1998, p. 4.

Kraemer, Kenneth L., and Jason Dedrick, "From Nationalism to Pragmatism: IT Policy in China," *Computer*, August 1995, pp. 64–73.

Lachica, Eduardo, "AlliedSignal Ends Plan to Coproduce Engines in China," *The Wall Street Journal*, 27 October 1995.

Lall, Sanjaya, "Technological Capabilities and Industrialization," *World Development*, Vol. 20, No. 2, 1992, pp. 165–186.

Lam, Willy Wo-lap, "Jiang Still Thinking It Through," *South China Morning Post*, 5 January 2000.

Laris, Michael, "Chinese Test Craft for Manned Orbits," *Washington Post*, 22 November 1999, pp. A1, A19.

Lawrence, Susan V., "Unfinished Business," *Far Eastern Economic Review*, 17 December 1998, p. 22.

Lawrence, Susan V., and Trish Saywell, "In Banks We Trust," *Far Eastern Economic Review*, 10 December 1998, pp. 58–60.

Layman, Patricia, "China and Biotechnology: Beginning of a Long March," *Chemical and Engineering News*, Vol. 74, No. 52, 23 December 1996, p. 13.

Lewis, John Wilson, and Xue Litai, *China's Strategic Seapower: The Politics of Force Modernization in the Nuclear Age*, Stanford, CA: Stanford University Press, 1994.

Lewis, Paul, "Forging New Bonds," *Flight International*, 6–12 November 1996, pp. 32–33.

Li Ning, "Moving Toward the Age of Information—The Rapid Development of China's PC Sector," *Beijing Review*, 4–10 November 1996, pp. 10–13.

Liu Shen, "Satellite and Terrestrial Microwave Communications in China," *IEEE Communications Magazine*, Vol. 31, No. 7 (July), 1993, pp. 38–40.

Liu, Sunray, "Microsoft Builds R&D Dream Team in Beijing," *EE Times*, 3 September 1999.

Lorell, Mark, *Troubled Partnership: A History of U.S.-Japan Collaboration on the FS-X Fighter* (MR-612/2-AF), Santa Monica, CA: RAND, 1995.

Lu Xinkui, "Tremendous Achievements, Brilliant Course—20 Years of Reform of China's Electronics Industry," *Renmin Ribao*, 5 January 1999, p. 12, in Foreign Broadcast Information Service, Daily Report: China (FBIS-CHI-99-009), 9 January 1999.

Macrae, Duncan, "Europe Celebrates Breakthrough on Chinese Market," *Interavia*, November 1996, pp. 26–28.

Mecham, Michael, "First Chinese-Built MD-83s Find U.S. Home with TWA," *Aviation Week and Space Technology*, 3 May 1993, p. 29.

Mecham, Michael, "Trunkliner Work Begins in China," *Aviation Week and Space Technology*, 4 September 1995a, p. 27.

Mecham, Michael, "China Learns Skills, Culture of Airframers," *Aviation Week and Space Technology*, 2 October 1995b, pp. 56–57.

Mecham, Michael, "Joint Ventures Star in Beijing," *Aviation Week and Space Technology*, 16 October 1995c, p. 22.

Mervis, Jeffrey, "The Long March to Topnotch Science," *Science*, Vol. 270, 17 November 1995, pp. 1134–1137.

Ministry of Science and Technology, People's Republic of China, *Zhongguo Kexue Jishu Zhibiao 1998 (China Science and Technology Indicators 1998)*, Beijing: Kexue Jishu Wenxian Chubanshe, 1999.

National Bureau of Statistics, People's Republic of China, *Zhongguo Tongji Nianjian (China Statistical Yearbook)*, No. 18, Beijing: China Statistics Press, 1999.

National Bureau of Statistics and Ministry of Science and Technology, *Zhongguo Keji Tongji Nianjian (China Statistical Yearbook on Science and Technology)*, Beijing: Zhongguo Tongji Chubanshe, 1999.

National Science Foundation, *Human Resources for Science and Technology: The Asian Region* (NSF 93-303), Washington D.C., 1993.

Nuclear Engineering International, "Datafile: China," October 1993, pp. 16–22.

Office of the Press Secretary, the White House, "Fact Sheet: Export Controls on Computers," 3 August 2000 (http://www.bxa.doc.gov/HPCs/WhiteHseFactSheetAug32000.html, accessed 29 October 2000).

Office of the Secretary of Defense, *Proliferation: Threat and Response*, Washington DC: U.S. Government Printing Office, November 1997.

Office of the Under Secretary of Defense for Acquisition and Technology (OUSD(A&T)), *Militarily Critical Technologies List (MTCL)*, Washington D.C., 1996.

Organization for Economic Cooperation and Development (OECD), *National Innovation Systems*, Paris, 1997.

Organization for Economic Cooperation and Development (OECD), *Main Science and Technology Indicators*, No. 1, Paris, 1999.

Paloczi-Horvath, George, "Chinese Fans," *The Engineer*, 22 May 1997, p. 22.

Pomfret, John, "Russians Help China Modernize Its Arsenal," *Washington Post*, 10 February 2000, pp. A17–A18.

Pratt & Whitney, "Three-Nation Aviation Joint Venture Launched in Xian, China," 9 October 1997 (http://www.prattwhitney.com/cgi-bin/browse-news.pl?AdminOp=shownews&newsdir=pw&newsfile =news.879536343, accessed 15 November 1998).

Pratt & Whitney, "Chengdu Joint Venture Underway," 1 April 1998 (http://www.prattwhitney.com/cgi-bin/browse-news.pl?AdminOp=shownews&newsdir=pw&newsfile=news.891458723, accessed 15 November 1998).

Qin Shijun, "High-Tech Industrialization in China," *Asian Survey*, Vol. XXXII, No. 12 (December), 1992, pp. 1124–1136.

Qiu Shouhuang, "Present Status and Outlook of Line Transmission Systems in China," *IEEE Communications Magazine*, Vol. 31, No. 7 (July), 1993, pp. 46–47.

Rehak, Alexandra, and John Wang, "On the Fast Track," *The China Business Review*, March–April 1996, pp. 8–13.

Roberts, Dexter, "The Race to Become China's Microsoft," *Business Week*, 18 November 1996a, pp. 62–63.

Roberts, Dexter, with Bruce Einhorn, "Going Toe to Toe with Big Blue and Compaq," *Business Week*, 14 April 1997, p. 58.

Roberts, Michael, "China: Cultivating Home-Grown Technology," *Chemical Week*, 15 May 1996b, p. 39.

Rotman, David, "Western Firms Look to Tap into China's R&D," *Chemical Week*, 30 August/6 September 1995, p. s10.

Rotman, David, with Andrew Wood, "Firms Look to Tap China's Research," *Chemical Week*, 31 August/7 September, 1994, p. s26.

Saywell, Trish, "Softly, Softly," *Far Eastern Economic Review*, 11 December 1997, p. 60.

Saywell, Trish, "Customized Genes," *Far Eastern Economic Review*, 7 May 1998, pp. 48–50.

Schoenberger, Karl, "Motorola Bets Big on China," *Fortune*, 27 May 1996, pp. 116–124.

Shen, Xiaobai, *The Chinese Road to High Technology: A Study of Telecommunications Switching Technology in the Economic Transition*, New York: St. Martin's Press, 1999.

Simon, Denis Fred, "Sparking the Electronics Industry," *The China Business Review*, January–February 1992, pp. 22–26.

Simon, Denis Fred, "From Cold to Hot," *The China Business Review*, November–December 1996, pp. 8–16.

Sokolski, Henry, "What Now for China Policy," *The Wall Street Journal*, 15 March 1999.

Solid State Technology, "China's 909 Project Will Seek World-Class Chipmaking Capabilities," June 1996, pp. 50, 54.

South China Morning Post, "Sky the Limit in Bid to Boost Home-Grown Technology," 1 November 2000.

State Statistical Bureau, People's Republic of China, *China Statistical Yearbook 1996*, Beijing: China Statistical Publishing House, 1996.

Summers, Robert, and Alan Heston, "The Penn World Table (Mark 5): An Expanded Set of International Comparisons, 1950–1988," *Quarterly Journal of Economics*, May 1991, pp. 327–368.

Sun Zhenhuan, *Zhongguo Guofang Jingji Jianshe* (*China's Economic Construction for National Defense*), Beijing: Junshi Kexue Chubanshe, 1991.

Suttmeier, Richard P., "China's High Technology: Programs, Problems, and Prospects," in Joint Economic Committee, ed., *China's Economic Dilemmas in the 1990s: The Problems of Reforms, Modernization, and Interdependence*, Washington D.C.: U.S. Government Printing Office, 1991, pp. 546–564.

Suttmeier, Richard P., "Emerging Innovation Networks and Changing Strategies for Industrial Technology in China: Some Observations," *Technology in Society*, Vol. 19, No. 3&4, 1997, pp. 305–323.

Suttmeier, Richard P., and Cong Cao, "China Faces the New Industrial Revolution: Achievement and Uncertainty in the Search for Research and Innovation Strategies," *Asian Perspective*, Vol. 23, No. 3, 1999.

Suttmeier, Richard P., and Peter C. Evans, "China Goes Nuclear," *The China Business Review*, September–October 1996, pp. 16–21.

Technology Review, "The Chinese Biotech Connection," 20 July 1992, pp. 19–20.

Tilley, Kevin J., and David J. Williams, "An Overview of the Chinese Electronics Industry," *1996 IEEE/CPMT International Electronics Manufacturing Technology Symposium*, pp. 329–336.

Tilley, Kevin J., and David J. Williams, "Technology and Change in the Chinese Electronics Industry," *IEEE Transactions on Components, Packaging, and Manufacturing Technology*, Part C, Vol. 20, No. 2 (April) 1997, pp. 142–151.

Todd, Iain, "China—the Inside Perspective," *Interavia*, February 1995, p. 18.

U.S. Arms Control and Disarmament Agency (ACDA), *World Military Expenditures and Arms Transfers 1996*, Washington D.C., 1997.

Wagner, Caroline S., *Techniques and Methods for Assessing the International Standing of U.S. Science* (MR-706.0-OSTP), Santa Monica, CA: RAND, October 1995.

Wang Chunyuan, *China's Space Industry and Its Strategy of International Cooperation,* Stanford University: Center for International Security and Arms Control, 1996.

Wang Zhongqiang, "China's Aero-Industry Hits the Road," *Interavia*, June 1995, pp. 17–20.

Wei Bian, "China's Nuclear Industry," *Beijing Review*, No. 25, 22–28 June 1998, pp. 12–14.

Williamson, Bonnie, "Not a Bang, Not Even a Whimper: U.S.-China Agreement in Effect," *Nucleonics Week*, 26 March 1998, p. 9.

Wolf, Charles, K. C. Yeh, Anil Bamezai, Donald Henry, and Michael Kennedy, *Long-Term Economic Trends, 1994–2015: The United States and Asia* (MR-627-OSD), Santa Monica, CA: RAND, 1995.

Wood, Andrew, and Ian Young, "China: The Year of Transition," *Chemical Week*, 28 August/4 September 1996, p. s1.

World Bank, *China 2020: Development Challenges in the New Century*, Washington D.C., 1997a.

World Bank, *World Development Indicators 1997*, Washington D.C., 1997b.

Wu, Yanrui, "Productivity Growth, Technological Progress, and Technical Efficiency Change in China: A Three-Sector Analysis," *Journal of Comparative Economics*, No. 21, 1995, pp. 207–229.

Xiao Qing, "Galaxy II, China's Space-Age Computer," *China Today*, Vol. 42, No. 3, 5 March 1993, pp. 42–43.

Xinhua, "China: Sino-British Joint Venture to Make Aero Engine Turbine Blades," 13 May 1997, in Foreign Broadcast Information Service, Daily Report: China (FBIS-CHI-97-133), 13 May 1997.

Xinhua, "Xian Rolls-Royce Venture Producing Aviation Engine Parts," 19 July 1998, in Foreign Broadcast Information Service, Daily Report: China (FBIS-CHI-98-200), 19 July 1998.

Xinhua, "China Puts Earth Resource Satellite into Orbit," 1 September 2000, in Foreign Broadcast Information Service, Daily Report: China (FBIS-CHI-2000-0901), 1 September 2000.

Yao Yan, Cao Zhigang, and Wang Jin, "R&D Activities on Wireless Systems in China," *IEEE Communications Magazine*, Vol. 31, No. 7 (July), 1993, pp. 42–45.

Ye Peida and Ren Xiaomin, "Status of R&D for Optical Fiber Communication Systems in China," *IEEE Communications Magazine*, Vol. 31, No. 7 (July), 1993, pp. 48–50.

Young, Ian, and Andrew Wood, "China: A Future Chemical Superpower," *Chemical Week*, 31 August/7 September 1994, pp. s1–s2.

Yuan Zhou, "Reform and Restructuring of China's Science and Technology System," in Denis Fred Simon, ed., *The Emerging Technological Trajectory of the Pacific Rim*, Armonk, NY: M. E. Sharpe, 1995, pp. 213–238.

Zhang Jinqiang, "Electronics Industry: Review and Prospects," *Beijing Review*, 18–24 March 1996, p. 22.

Zhou Bingkun, "An Overview of Optical Device Research in China," *IEEE Communications Magazine*, Vol. 31, No. 7 (July), 1993, pp. 62–65.

Zhou Guangzhao, "The Course of Reform at the Chinese Academy of Sciences," *Science*, Vol. 270, 17 November 1995, p. 1153.

Zhu Yilin, "Fast-Track Development of Space Technology in China," *Space Policy*, May 1996, pp. 139–142.